> To the gardeners and creators who find peace in the soil and beauty in the simple things. This book is for you, for the roots you plant, and the dreams you nurture. May your journey with nature be as fulfilling as the blooms you grow.

In every seed lies the promise of beauty, and in every garden, the power to heal

Copyright © April 2025 by Cailin Cooper
All rights reserved.

No part of this book may be reproduced, distributed, or transmitted in any form or by any means, including photocopying, recording, or other electronic or mechanical methods, without the prior written permission of the author, except in the case of brief quotations embodied in critical reviews and certain other noncommercial uses permitted by copyright law. For permission requests, write to the publisher, addressed "Attention: Permissions Coordinator," at the address below.

Cooper Delivered
info@cooperdelivered.com.au

First Edition: April 2025

Disclaimer

Seeds and Growing Practices

The seeds and growing practices described in this book are intended for educational purposes only. While every effort has been made to ensure the accuracy of the information provided, it is important to note that growing conditions can vary significantly based on factors such as climate, soil type, and regional pests. Readers are encouraged to research and follow best practices for their specific location. Additionally, some plants may be invasive or restricted in certain areas, so please consult local regulations before planting. Readers are also encouraged to source their seeds and plants ethically, supporting sustainable and fair-trade practices.

Recipes and Natural Remedies

The recipes and natural remedies included in this book are based on traditional and historical uses of plants for beauty and wellness. These recipes are intended for external use only, unless otherwise specified. While many of the ingredients are considered safe for most individuals, it is important to conduct a patch test before using any new product, especially if you have sensitive skin or allergies. The traditional uses of plants mentioned in this book are based on cultural practices that should be respected and acknowledged. Readers are encouraged to honor the cultural heritage of these remedies.

Safety Precautions

Allergies: Always check for potential allergies to any ingredients before use. If you are unsure, consult with a healthcare professional.

Medical Conditions: If you have any medical conditions or are pregnant, nursing, or taking medication, please consult a healthcare provider before using any natural remedies or herbal products.

Proper Storage: Store homemade products in clean, airtight containers, and use them within recommended time frames to ensure safety and effectiveness.

Child Safety: Keep all ingredients, seeds, and products out of reach of children and pets.

Legal Considerations: Please be aware of the legal status of certain herbs and plants in your region, as some may be regulated or restricted.

Sustainability Note

Readers are encouraged to practice sustainable gardening methods and consider the environmental impact of their gardening choices. Supporting eco-friendly practices contributes to the well-being of our planet.

Liability

The author and publisher of this book disclaim any liability for any adverse effects or consequences arising from the use of the seeds, recipes, or information contained in this book. Readers assume full responsibility for the use of any plants, seeds, or recipes mentioned herein. By using the information provided in this book, readers agree to do so at their own risk and to take full responsibility for any actions taken based on the content of this book.

Why This Book Was Born

There was a moment—quiet, unassuming—when I held a handful of damp soil and felt the Earth whisper to me. It wasn't a shout or a sign. It was a gentle knowing. A reminder that everything we need has already been given to us.
This book was born from that moment.
I created Soil to Roots for the barefoot gardeners and balcony growers, the potion makers and dream chasers, the ones who believe magic exists in both compost bins and calendula petals. I wrote it for the women craving reconnection—not just to nature, but to themselves.
Because I had forgotten, too.
I had forgotten how sacred it is to dig with my hands. How powerful it is to grow a plant from seed and watch it bloom. How healing it feels to create a shampoo bar or salve with something I nurtured from the ground up. I remembered that beauty isn't something bought—it's something grown, harvested, stirred, and shared.
This book is a blend of the practical and the poetic. A field guide for beginners. A balm for those burned out by modern beauty. A woven basket of stories, cultural memory, and recipes rooted in Earth wisdom. And most of all—it's an invitation.

To slow down.
To grow with intention.
To rediscover that you, too, are part of this living ecosystem.
We don't garden to escape life. We garden to remember how to live it.
So welcome to Soil to Roots.
Let's plant something beautiful together.

Cailin Cooper

How to Use This Book

You don't need to be a gardener. You don't need a yard. You don't need to know the difference between rosemary and lavender. All you need is a willingness to reconnect—with your hands, with your heart, and with the Earth beneath your feet.

Here's how to make the most of this journey:

- Start where you are. If you're a complete beginner, skip to the "Where to Begin (Even With Nothing)" chapter. If you've already got herbs on your windowsill, jump straight into the plant profiles and recipes.
- Take your time. This book is designed to be savored like a cup of herbal tea. You can read it cover to cover or open it to the chapter that calls to you most.
- Get your hands dirty. The goal is not just to read—but to do. Plant something. Make something. Get messy. This is a workbook for your hands and your spirit.
- Use it as a seasonal guide. The tips and planting calendar can help you align with the seasons and local weather. Grow slowly, sustainably, and with joy.
- Come back often. You'll grow, and so will your garden. What you needed to read now may change in six months. That's the beauty of this life—it's always unfolding.

So light a candle. Brew a cup of something warm. Let's begin the journey back to our roots

ROOTED IN PURPOSE The Philosophy of Growing

Why We Garden, Why It Matters

To garden is not simply to grow—it is to return.
To remember.
To reconnect.
In the act of placing your hands into the soil, you are not just planting seeds—you are grounding yourself in something far greater than you. Gardening is one of the oldest human rituals. It predates empire, language, and borders. Before there were tools, before there were rules, there were hands and seeds and the earth between them.
We garden because we are wired to co-create with nature.
We garden because it slows the spinning world.
In a culture obsessed with instant gratification, gardening teaches us patience. In a time of disconnection, it anchors us. In a world of concrete and screens, it offers us the most ancient form of therapy: dirt under the nails and wind on our skin.
But beyond the joy of harvests and blooms, there is something deeper—a whisper of something sacred. To tend a garden is to honour the cycle of life. Every wilted leaf, every composted root, becomes the foundation for something new. There is no waste in nature. Only transformation.

A Garden Is a Mirror

Your garden will reflect you—not just your choices in flowers or herbs, but your rhythms, your energy, even your fears. The more you show up for it, the more it shows up for you. It is your teacher, your healer, your silent companion.
Some days it thrives. Some days it falters.
Just like you.
And in learning to care for it, you will learn to care for yourself.

Cultural Wisdom from the Ground Up

Across cultures and centuries, the act of growing herbs, vegetables, and flowers has never been just about putting food on the table. It has always been deeper—an expression of identity, a connection to the divine, a way to nurture both body and spirit. Gardens were never just for sustenance. They were for ceremony. For healing. For beauty. For power.
In Ireland, herbal gardens flourished not only for medicine, but for protection and personal care. Sprigs of rosemary weren't just tossed into stews—they were tucked beneath pillows to encourage dreams and rubbed into the scalp to stimulate growth and memory. Chamomile was brewed into hair rinses that

brightened golden locks and soothed the senses, while nettle was the trusted herb of strength—for the body, the blood, and the strands on your head.

In Ballardong Noongar country, wisdom came not from books but from the land itself. Ochre and clay weren't just tools of survival—they were mediums of beauty, medicine, and identity. The seasons taught when to rest, when to gather, when to plant, and when to leave the earth untouched. Hair and skin were cared for with sacred, earth-sourced ingredients, passed down in songlines and story.

In India, henna was not just body art—it was medicine, self-care, adornment, and protection. Used to dye hair, cool the skin, and prepare for weddings and rites of passage, it was mixed with oils and herbs for even deeper benefits. Tulsi, or holy basil, was used in baths, teas, and facial water to cleanse the aura and calm the skin. Turmeric became a symbol of brightness—both spiritual and physical—used in masks and scrubs to illuminate the face.

In Morocco, women gathered in traditional hammams—ritual bathhouses—to use rhassoul clay, harvested from beneath the Atlas Mountains. Rich in minerals, this clay cleansed and softened the skin, preparing women for prayer, celebration, or simply everyday life. Argan oil, known as 'liquid gold," was pressed by hand in women's cooperatives, then smoothed onto skin and hair as a deep act of nourishment.

In China, beauty and health were one and the same. Peony root, goji berries, ginseng, and green tea were all grown and consumed not just for vitality but for skin clarity and graceful aging. Rice water became a beauty tonic for glossy hair and glowing skin, and jade tools were used in facial massage to release tension, smooth fine lines, and draw in cooling qi.

Across Africa, plants like shea, aloe vera, and baobab were not commercial products. They were gifts from the earth, gathered, pressed, and applied by generations of women who used their gardens to care for their families, their bodies, and their spirits. Hair oils, body butters, salves, and masks—made from simple plants—carried centuries of tradition in every drop.

These rituals weren't born in laboratories or factories. They were born in gardens, in kitchens, in ritual spaces under open skies. These weren't just beauty secrets—they were legacies. Acts of sovereignty. Moments of power that reminded women everywhere: you are sacred, and the Earth provides for you.

So when you plant calendula in your garden, know that you are stepping into this lineage.

When you brew your own lavender rinse or crush rosemary into oil for your hair, know that you are not just making a remedy.

You are reclaiming a birthright.

You don't need a massive plot of land to be part of this.

You just need intention. A pot. A window. A bit of sunlight.

And a willingness to remember what we were never meant to forget:
- That healing and beauty are grown, not bought.
- That nature will always give us what we need—if we learn to listen.
- And that your garden can be your pharmacy, your vanity, your sanctuary, and your revolution—all at once.

You are not just a gardener.
You are a keeper of rituals.
A heir to wisdom.
A maker of magic—from soil to root, from root to bloom.

Why It Still Matters Today

In our modern world, gardening can feel like an outdated luxury. Who has the time to tend herbs when there are emails to send, bills to pay, and deadlines looming? But here's the truth:
Gardening is not a hobby. It's a quiet revolution.
When you grow a basil plant on your windowsill or nurse a tomato vine in your courtyard, you reclaim something the world is constantly trying to take from you—presence. Agency. A sense of enoughness.
We are bombarded with messages that healing comes in bottles, beauty in boxes, peace in a paywall. But the truth is, it can come from a handful of mint. From a lavender balm you made with your own two hands. From running your fingers through thyme and breathing in resilience.
Gardening isn't just practical—it's deeply political.
It resists waste. It decentralizes power.
It says: "I can care for myself. I can nourish my family. I can create beauty without asking permission."
And more than that—it slows you down. When you pause to water, to prune, to harvest, your nervous system softens. Your breath lengthens. Your thoughts quiet.
In the garden, you're not behind.
You're not too much.
You're not failing.
You are simply—here.

So when you ask:
- What nourishes me?
- What do I want to create?
- What do I want to pass on?

Remember that the answers might not come in a lightning bolt. They might

come in the way your hands move through soil. In the satisfaction of the first bud. In the softening of your shoulders after a day outside.
Because gardening isn't about perfection.
It's about participation.
It's about remembering that even one seed, placed with intention, is enough. Enough to begin. Enough to heal. Enough to change the story.

In Every Seed Lies a Thousand Forests

Connection, Potential & Earth Magic

Every time you hold a seed in your hand, you are holding history. You are holding memory. You are holding potential.

A seed may look small—unassuming, even. Dry. Ordinary. But inside it is a blueprint older than time. A single calendula seed carries the wisdom of generations of gardeners, herbalists, healers, and women who once sat in moonlight, crushing its petals into oil to soothe dry skin or bless a newborn's first bath. A basil seed holds centuries of culinary joy and sacred temple offerings. Lavender, rosemary, thyme—each tiny beginning holds a forest of stories.

And when you plant that seed? You're not just growing a plant.
You are becoming a creator.
A healer.
A guardian of life.

Seeds Connect Us

Across all traditions, seeds were sacred. In Aboriginal Ballardong culture, seeds from native grasses weren't just harvested—they were listened to. Women observed the season's rhythms before gathering them, grinding them into nourishing flours or saving them for sacred smoke ceremonies. In ancient Egypt, black cumin—known today as "the seed of blessing"—was buried with pharaohs for the afterlife. And in the Andes, quinoa was planted in ritual, with the first seeds sown by high priestesses.

Even today, gardeners in every corner of the world hold seed-saving ceremonies, trade heirloom varieties, and pass down stories alongside their seeds. To grow from seed is to join a global conversation without words.

Why Seeds Matter in a Beauty Garden

When we talk about growing for self-care, for rituals, for recipes, we start with seeds because they teach us patience, presence, and intention. A seedling doesn't rush. It doesn't bloom before its time. It doesn't worry if the sun will return. It simply grows—slowly, confidently, quietly.

When you grow from seed, especially herbs and flowers for natural beauty recipes, you begin a cycle of deep connection:
- You witness birth, death, and rebirth on your windowsill.
- You begin to trust your hands to nurture life.
- You start creating products with soul—infused oils, masks, tinctures, and teas that are alive with your energy.

This isn't about doing it perfectly. It's about doing it purposefully

Start Small, Dream Big
One seed can become a handful of petals. One rosemary plant can perfume your entire garden and your hair oil. One tiny sprout of thyme can become the heart of your healing salve. You don't need acres of land. You need:
- A pot of soil
- A patch of sun
- A bit of belief in your own hands

Start with the seeds you love. Start with what you use. Think of what you'd love in your natural hair rinse, your foot soak, or your calming balm. That's your starting point.

There Is Magic in the Act of Planting
When you plant a seed, you're saying:
"I believe in tomorrow.
I believe in slow miracles.
I believe I can create something beautiful from almost nothing."
The garden doesn't ask for credentials. It doesn't care what you know. It simply invites you to begin. And each seed you tuck into the soil becomes a love letter to the Earth—and to yourself.
So start where you are. Whether you grow one pot of mint on your windowsill or turn your backyard into a healing apothecary, your effort echoes across time. The ancestors who tended gardens in clay jars, forest clearings, and desert courtyards are nodding in approval.
Because they know—what you grow may look small now.
But in every seed, there lies a thousand forests.
In every gardener, there lies a thousand generations.
And in you? There lies the beginning of something truly powerful.

Where to Begin (Even With Nothing)

Growing in Small Spaces, on a Budget, or in a Tiny Yard
You don't need acres. You don't need raised beds. You don't even need fancy tools or the perfect soil.

You just need a bit of space, a touch of curiosity, and the belief that something can grow from very little.

Some of the most magical gardens start in forgotten corners—windowsills, recycled pots, sunny balconies, and cracked teacups. They begin not with a shopping spree but with scraps from the kitchen, seeds gifted by a neighbour, or cuttings passed down from someone who believed in your green thumb before you did.

This is your invitation to begin exactly where you are, with what you have.

You Don't Need More. You Just Need to Begin.

Let's debunk the biggest myth right away:
You don't need a lot of money to start a garden.
What you need is intention—and a little guidance.
Start with what's already around you. An old bucket becomes a planter. A cracked mug holds basil by the sink. A cardboard egg carton becomes a seedling tray. Soil can be sourced for free or made from scraps. Herbs can be grown from cuttings. You are not behind. You are not unprepared. You are just beginning.
Tip: Look around your home right now.
Can you spot three containers that could hold soil? That's your garden.

The Four Basic Things Every Garden Needs (Even in a Tiny Space)

Before you plant your first seed, understand what plants really need. Whether you're gardening on a balcony, in a windowsill, or a shared yard, these are your non-negotiables:

<u>Light</u> - Most herbs and flowers need 4–6 hours of sunlight per day. South-facing windows are ideal. No sun? Try shade-loving herbs like mint, parsley, or lemon balm.

<u>Water</u> - Plants don't need much, but they do need consistency. A reused spray bottle works for delicate sprouts, and water from rinsing rice or veggies adds nutrients.

<u>Soil</u> - Start with basic potting mix. You can improve it over time with compost, coffee grounds, or crushed eggshells. Don't worry about perfection—plants want to grow.

<u>Love</u> - Yes, really. Check in daily. Touch the soil. Whisper if you must. Presence grows plants as much as sunlight does.

Budget Gardening Hacks: Grow More, Spend Less

Seed Swap Groups: Local libraries, community centres, and even Facebook groups often host free seed exchanges.

Kitchen Scraps That Regrow: Green onions, celery, basil, lettuce, and garlic can all regrow in water before moving to soil.

Cuttings from Friends: A rosemary clipping, a mint runner, or a stem of lavender can root in water and become your first plant.

Recycled Pots: Yogurt tubs, tin cans, egg cartons—poke holes in the bottom and you're ready to plant.

Composting Basics: Even a small bin or bucket can transform food scraps into rich soil (banana peels, coffee grounds, veggie bits).

Choosing What to Grow First

Start with what you'll use and what thrives easily. Think low-maintenance, high-reward:

Basil - Fast-growing, loves warmth, great for hair rinses & food.

Rosemary - Hardy, boosts scalp health, magical in salves.

Calendula - Beautiful petals for healing oils, great in any garden.

Lavender - Aromatic, calming, loved by bees, perfect for sachets.

Mint - Grows fast (almost too fast!), cooling, great in tea.

Chamomile - Delicate flower with soothing magic.

Start with one or two. Master them. Love them. Then expand. You don't need a forest to begin. You just need a single green shoot reaching toward the sun.

Small-Space Setups That Work

Windowsill Garden: Line up pots on the sunniest ledge in your home. Herbs do beautifully here.

Vertical Garden: Use a wooden pallet or hanging shoe organizer to create a wall of herbs.

Balcony Boxes: Attach long troughs to railings for a balcony oasis.

Shelf Garden: Stack shelves near a window—each level gets sun.

Indoor Jars & Glasses: Transparent jars let you watch roots form—especially magical for kids and curious adults alike.

From Sprout to Self-Care

Growing for beauty and well-being starts now—not once you have "the perfect setup."

Even in your first season, you can:

Harvest rosemary for a hair rinse that awakens the scalp.

Steep lavender in oil for a sleep balm.

Dry mint leaves for foot soaks or refreshing sprays.

And all of that can begin in a recycled jam jar on your windowsill.

Why Growing Anything Is Revolutionary
In a world that moves too fast, that sells you beauty in bottles and wellness in plastic tubs, growing your own is an act of rebellion.
You are saying:
"I trust my hands."
"I believe in patience."
"I will grow something healing, something sacred, something mine."
You are stepping away from consumption and returning to creation.

You Don't Have to Have It All Figured Out
If all you have today is a packet of seeds and an old pot, that's enough. If all you can do is place your hands in soil and whisper a promise, *that is enough*. The garden is always ready to meet you where you are.
The Earth is generous.
She does not require perfection.
She only asks you to begin.
And so—begin.

The Heart of the Garden

Beneath every beautiful bloom, lush leaf, or thriving herbal patch is something most people never see: the quiet, living foundation of a garden. It's not glamorous. It's not loud. But it's everything. If you're just starting out, this is where your journey truly begins—not with the plant, but with the conditions that help it thrive.
Because a plant is only as strong as the roots it grows. And roots are only as strong as the soil, the light, the water, and the rhythm of the seasons that hold them.
Let's get your garden growing.

Soil: The Living Foundation
Think of soil not as "dirt," but as a living, breathing ecosystem. It holds microbes, fungi, nutrients, and minerals that your plants rely on. Healthy soil is soft, rich, and dark—like chocolate cake. Poor soil, on the other hand, can be compacted, sandy, or dry.
Tips for Getting Great Soil (Even on a Budget):
Start with what you have: Grab a handful. Does it clump but crumble when pressed? That's good. Is it hard as a brick or full of sand? You'll need to amend it.
Add compost: You can make your own or buy a bag from a local nursery. Compost brings life back into dead soil.

Mix in worm castings or organic matter: These feed the soil's micro-ecosystem and help moisture retention.
Consider a raised bed or potting mix: If your yard soil is too hard to fix immediately, grow in containers using high-quality soil blends.

Indigenous Gardening Insight
Many First Nations people across the globe, including the Ballardong Noongar in Western Australia, have gardened in sync with the land's rhythms. Instead of relying solely on commercial planting calendars, they observed animals, flowering patterns, and rainfall to know when to plant.
Suggestion: Research your region's Indigenous seasonal calendar. It can guide your planting more intuitively than any app.

Sun: The Source of Strength
Plants need sunlight to photosynthesize, which means turning light into food. But not all plants love the same kind of light.
Know Your Plant's Sun Needs:
Full Sun = 6+ hours of direct sunlight
(Lavender, rosemary, calendula, tomatoes)
Partial Sun/Partial Shade = 3–6 hours
(Mint, chamomile, parsley, nasturtium)
Full Shade = Less than 3 hours
(Ferns, lemon balm, violets)
Pro Tip: Track your garden space with a notebook. Observe how much light different areas get throughout the day. Most beginners overestimate their sun.

Water: A Delicate Balance
Too much water drowns roots. Too little, and your plant goes limp. Watering is an art of balance, not just a chore.
Watering Wisdom:
Check the soil: Stick your finger 2 cm into the soil. If it's dry, it's time to water.
Water in the early morning: Plants absorb water best before the sun gets intense.
Mulch, mulch, mulch: Add a layer of straw, bark, or leaves on top of your soil. It keeps moisture in and reduces the need for frequent watering.
Collect rainwater: It's free, gentle on plants, and environmentally conscious.

Seasons: The Earth's Rhythm
Every region has its rhythm. The trick is not to fight it, but to garden in harmony with it.
Seasonal Guide for Beginners:
Spring = sowing seeds, planting leafy herbs and fast growers.

Summer = maintain, water deeply, harvest often.
Autumn = plant roots, garlic, onions, and perennials. Collect seeds.
Winter = rest, compost, plan, and nourish your soil.
If you're in Australia or working with native plants, seasonal planting may not follow a four-season model. The Noongar calendar, for example, has six seasons—each one deeply attuned to signs in the environment like flowering gums or emu breeding calls.

A Quick Reference for Success

Keep this checklist handy:

Have I enriched my soil?
Do I understand my garden's sunlight exposure?
Am I watering the right amount and at the right time?
Do I know what grows best in this season?

The Heartbeat of a Garden
You don't need to have it all figured out. You just need to begin.
The earth will meet you halfway, if you're willing to show up. Over time, the soil softens, the seeds sprout, and your confidence grows along with the plants.
And as you learn the needs of each flower, root, and herb, you'll notice something else shifting: your pace. Your breath. Your own inner rhythm aligning with the Earth.
This is the true gift of gardening—not perfection, but presence.

Calendula: The Golden Healer

Botanical Name: Calendula officinalis
Common Names: Pot Marigold, Gold Bloom, Mary's Gold
Element: Fire
Planetary Ruler: Sun
Season: Spring to Autumn
Energy (Magical/Medicinal): Uplifting, Healing, Protective
Companion Plants: Tomatoes, Carrots, Basil
Beneficial Neighbors: Cucumbers, Peas
Avoid Planting With: None known—Calendula is a friendly companion
Indoor or Outdoor Growing: Best outdoors (needs full sun), but can be grown indoors near a bright window or under grow lights

The Soul of Calendula

Calendula has long been seen as the sunshine of the garden—a flower that not only blooms with brilliant gold but also radiates healing, warmth, and protection. In ancient cultures, it was planted to ward off illness, woven into crowns for ritual ceremonies, and infused in oils by midwives and herbalists alike.

In European folk magic, calendula was a symbol of constancy and remembrance. In Mexican traditions, it's deeply linked to Día de los Muertos, where petals guide spirits home. In many Indigenous and earth-centered cultures, its golden hue was said to bring clarity, and its blooms were seen as messengers from the sun.

How to Grow Calendula

Location: Full sun to part shade
Soil: Well-draining, slightly sandy soil (though calendula is forgiving and thrives even in average soil)

Planting:
You can direct sow seeds straight into the ground in early spring as soon as the soil is workable.
Alternatively, start seeds indoors 4–6 weeks before the last frost in seed trays or biodegradable pots for a head start.
No special germination treatment is needed—calendula germinates easily in warm, moist soil.
Watering: Regular watering until established, then reduce. Calendula is drought-tolerant but appreciates consistent moisture.

Propagation from Cuttings: Calendula is typically grown from seed, not cuttings. While softwood cuttings can technically root in moist soil, they're often weak. Seed propagation is the most reliable and vigorous method.

Tip: Deadhead often to prolong blooming. If you let some flowers go to seed, calendula will self-sow generously for the following season—nature's own way of gifting you a golden harvest.

Garden Wisdom:
Calendula is self-seeding. If you let a few blooms go to seed, you'll have surprise flowers next year—an offering from the Earth itself.

Magical & Healing Uses
Spiritual Protection: Scatter petals around doorways or windows to invite warmth and block negative energy.
Emotional Support: A calendula tea or foot soak is soothing for those experiencing burnout or sadness.
Skin Healing: Calendula is known for its ability to heal wounds, rashes, and dryness. It brings softness and strength.

Beauty from the Garden
Golden Glow Infused Oil
Use this oil in your skincare routine for soothing, regenerative beauty rituals.
You'll Need:
- 1 cup dried calendula petals
- 1 cup cold-pressed olive oil or sweet almond oil
- A clean glass jar

Instructions:
- Add calendula to the jar and pour the oil over it.
- Let sit in a sunny window for 2–4 weeks, shaking daily.
- Strain and store in a dark bottle.

Use: As a gentle moisturizer, makeup remover, or to heal sun-kissed skin.

Harvesting & Storage
Harvest: Pick flowers when fully open, ideally mid-morning after dew has dried.
Drying: Lay flat in a single layer on a screen or basket in a warm, dry place out of direct sunlight.
Storage: Store in airtight jars away from light and moisture. Petals should retain their color and scent.

CHAMOMILE

THE GENTLE SOOTHER

Chamomile

Botanical Name: Matricaria chamomilla (German Chamomile) / Chamaemelum nobile (Roman Chamomile)
Common Names: Ground Apple, Whig Plant, Maythen
Element: Water
Planetary Ruler: Moon
Season: Spring to Late Summer
Energy (Magical/Medicinal): Calming, Soothing, Protective, Dream-enhancing
Companion Plants: Cabbage, Onions, Mint, Beans
Beneficial Neighbors: Brassicas, Cucumbers, and other herbs (improves their health)
Avoid Planting With: None strongly known—chamomile is generally companion-friendly
Indoor or Outdoor Growing: Can be grown both indoors (with sunlight or grow lights) and outdoors; prefers full sun and well-drained soil

Spiritual & Cultural Roots

Chamomile has long been known as the plant of peace, favored in ancient Egypt, Rome, and among Celtic healers. Egyptians dedicated it to the Sun God Ra, using it in beauty rituals and for treating fevers. In European folklore, it was planted in gardens for protection and to ward off hexes. Its name comes from the Greek chamaimēlon—"earth apple"—because of its soft, sweet scent.

Today, it's still a staple in herbal medicine, and it whispers to us through sleep teas, calming baths, and golden tonics.

Magical & Healing Properties

Energetically soothing – calms nerves and emotional tension
Associated with the moon, water, and feminine energy
Spiritually used for: peace, protection, dreamwork, and healing rituals
Beauty benefits: brightens skin, lightens hair, reduces redness, and soothes inflammation.

How to Grow Chamomile

Types: There are two common types:
German Chamomile (Matricaria chamomilla): an annual that self-seeds easily and grows upright.
Roman Chamomile (Chamaemelum nobile): a low-growing perennial that forms a carpet-like groundcover.
Location: Full sun preferred, though it can tolerate some shade.
Soil: Light, well-drained soil—avoid overly rich soils or it may produce more

foliage than flowers.

Planting:
Direct sow seeds into the garden in early spring after the last frost.
Chamomile seeds are very small and need light to germinate, so press them gently into the soil rather than covering them.
Keep the soil moist until seedlings appear (usually within 7–14 days).

Propagation from Cuttings: Roman chamomile can be propagated by division or softwood cuttings. German chamomile, however, is best grown from seed.

Watering: Water moderately. Chamomile prefers slightly dry soil over wet feet.

Harvesting: Pick flowers when they're fully open, usually in early morning after dew has dried. For best potency, harvest continuously to encourage more blooms.

Beauty Recipe: Soothing Chamomile Eye Compress
For tired eyes, puffiness, and dark circles.
- 1 tsp dried chamomile flowers
- ½ cup hot water
- 2 cotton rounds or cloths

How to Use:
Steep the chamomile in hot water for 5–10 minutes. Let cool slightly, then soak cotton rounds. Place over closed eyes for 10–15 minutes.
Whisper: "I soften. I soothe. I see clearly."

Tip: Companion Planting Bonus
Chamomile is known as the "plant doctor" in gardens—it helps neighboring herbs and vegetables grow stronger and resist disease. It especially benefits mint, basil, cabbage, and cucumbers. It's also a magnet for pollinators, including bees and beneficial hoverflies.

LAVENDER

THE SACRED BALM OF SPIRIT & SKIN

Lavender

Botanical Name: Lavandula angustifolia
Common Names: Lavender, English Lavender, True Lavender
Element: Air
Planetary Ruler: Mercury
Season: Summer
Energy: Calming, Cleansing, Protective
Companion Plants: Sage, Thyme, Rosemary
Beneficial Neighbors: Cabbage, Carrots, and Roses
Avoid Planting With: Mint (due to differing water needs)
Indoor or Outdoor: Best grown outdoors in full sun, but can be grown indoors in a large container with excellent drainage and strong sunlight.

Spiritual & Cultural Roots

Lavender has been cherished for over 2,500 years—as a sacred scent, a healer's balm, and a symbol of peace and purity. Ancient Egyptians used it in mummification and perfumes. The Greeks bathed in it. In medieval Europe, it was tucked into linen drawers and under pillows to ward off evil and calm nightmares. Romans believed it purified the soul, and in folklore, lavender was woven into protection charms for women and children.

Lavender represents clarity, calm, and sacred femininity. It is said to restore emotional balance and help open the third eye for intuitive insight.

Magical & Healing Properties

Spiritually used for: cleansing, peace, restful sleep, protection, and connection to divine feminine energy
Energetic associations: air, mercury, the third eye chakra
Beauty benefits: balances oil production, soothes inflammation, helps heal acne and redness, calms scalp irritation

How to Grow Lavender

Type: Choose varieties like English Lavender (Lavandula angustifolia) for culinary and beauty uses.
Location: Full sun—lavender loves light!
Soil: Well-draining, sandy or loamy soil. Slightly alkaline (add a bit of crushed eggshell or garden lime if your soil is acidic).
Planting: Lavender is best started from cuttings or small plants, as seeds can be slow to germinate.
If starting from seed, use a heat mat and patience, as it can take 2–4 weeks to sprout.

Space plants about 30–45cm apart for good airflow.

Propagation from Cuttings:
Take softwood cuttings in spring or semi-hardwood cuttings in late summer.
Dip the ends in cinnamon or rooting hormone and place in damp sand or perlite.

Watering: Let the soil dry between watering. Overwatering is the most common lavender mistake.

Harvesting: Harvest flower spikes when buds are fully formed but not yet open for best aroma and potency.

Beauty Recipe: Lavender-Rose Facial Steam

For calming the skin and mind.
- 1 tbsp dried lavender
- 1 tbsp dried rose petals
- A bowl of hot water and a towel

How to Use:
Place herbs in a bowl, pour boiling water over them, and place your face 8–10 inches above the bowl. Drape a towel over your head to trap steam. Breathe deeply for 5–10 minutes.

Whisper: "I return to peace. I let softness lead."

Tip: Lavender for the Hair
Lavender balances oil production on the scalp, promotes hair growth, and soothes flakiness.
Make a lavender scalp rinse with cooled lavender tea and a splash of apple cider vinegar after shampooing.

Bonus Magic: Lavender in the Garden

Lavender is a pollinator magnet. Bees, butterflies, and beneficial insects adore it. Plant it near vegetables to attract allies and repel pests. It also grows well with rosemary, thyme, and sage—your goddess garden dream team.

Aloe Vera

Botanical Name: Aloe barbadensis miller
Common Names: Aloe, Burn Plant, First Aid Plant
Element: Water
Planetary Ruler: Moon
Season: Warm seasons (perennial in warm climates)
Energy: Cooling, Healing, Protective
Companion Plants: Basil, Lavender, Rosemary (similar growing needs)
Beneficial Neighbors: Succulents, Cacti
Avoid Planting With: Plants needing frequent watering or poor drainage
Indoor or Outdoor: Ideal as an indoor plant; outdoors in warmer, frost-free climates.

Spiritual & Cultural Roots

Aloe has been known for millennia as the "plant of immortality." Used in Ancient Egypt, Mesopotamia, India, and Africa, it's long been treasured for its cooling, healing, and regenerative powers. Cleopatra herself was said to use aloe as part of her beauty ritual, while in Ayurvedic medicine, aloe is used to balance Pitta dosha (heat, inflammation, and digestion).
In many traditions, aloe is also protective—planted at the door to keep away evil or hung over homes to bless them with peace and healing. It symbolizes renewal, longevity, and gentle strength.

Magical & Healing Properties

Spiritually used for: healing, renewal, emotional balance, protection
Energetic associations: water, moon, sacral and heart chakras
Beauty benefits: hydrates, soothes burns and inflammation, helps with acne, scars, and dryness

How to Grow Aloe

Type: Aloe vera (Aloe barbadensis miller) is the most commonly used medicinal type.
Location: Needs bright indirect light or filtered sun. Can tolerate full sun but may scorch in very hot climates.
Soil: Well-draining cactus or succulent mix. Mix sand, perlite, or pumice into your soil for drainage.

Planting:
Best grown from pups (baby offshoots at the base of the mother plant) rather than seeds.

If using pups:
Gently remove the pup with some root attached.
Let it dry for 24 hours to allow the wound to callous.
Plant in dry soil and wait a week before watering.

Watering:
Water deeply, but infrequently. Let soil dry out completely between waterings.
Water less in winter. Overwatering is the #1 killer of aloe.

Harvesting:
Snip mature outer leaves close to the base. Let any sap (yellow latex) drain off, then use the clear inner gel.

Beauty Recipe: Aloe Glow Gel
For soothing, hydrating, and smoothing the skin.
- 1 fresh aloe leaf (filleted to collect the clear gel)
- 2 drops lavender essential oil
- 1 drop vitamin E oil

How to Use:
Blend all ingredients. Store in a small glass jar in the fridge for up to a week. Apply a small amount to clean skin after sun exposure, shaving, or as a nightly serum.

Bonus: Aloe Hair & Scalp Mask
Aloe deeply hydrates and calms the scalp, promoting healthy growth and shine.
- Mix 2 tbsp aloe gel with 1 tbsp coconut oil and a few drops of rosemary oil.
- Massage into scalp and ends. Leave for 20–30 mins, then rinse and shampoo.

Tip: Aloe for Emotional Healing
Place an aloe plant in your space to absorb emotional heat, especially after arguments, grief, or burnout. Its soft energy cools and nurtures, inviting restoration. Whisper to it: "Help me soften, help me begin again."

Can It Be Grown from Seed or Cutting?
Seed: Yes, but very slow and not ideal for beginners. Germination may take 3–4 weeks, and plants can take years to mature.

Cuttings: Not effective—aloe doesn't root from leaf cuttings like other succulents.

Pups are best—look for healthy offsets from a mature plant for the easiest success.

Rosemary

Botanical Name: Rosmarinus officinalis
Common Names: Rosemary, Dew of the Sea
Element: Fire
Planetary Ruler: Sun
Season: Evergreen; harvest year-round in warmer climates
Energy: Stimulating, Protective, Clarifying
Companion Plants: Cabbage, Beans, Carrots, Sage
Beneficial Neighbors: Repels pests from sensitive plants
Avoid Planting With: Basil (different water needs)
Indoor or Outdoor: Can be grown indoors with good airflow and sun; prefers outdoor gardens in full sun.

Spiritual & Cultural Roots

Rosemary has long been a sacred plant of memory, purification, and love. In Ancient Greece, students would wear rosemary crowns to boost focus and memory during exams. In Rome, it symbolized loyalty and was woven into wedding garlands and burial rites.

Among many cultures, rosemary is burned as a smoke cleanse to ward off negativity. In witchcraft and folk medicine, it's known as a plant of protection, clarity, and inner fire.

This herb isn't just for cooking—it's a time-honored beauty tonic and ritual plant, bridging body, mind, and spirit.

Magical & Healing Properties

Spiritually used for: memory, purification, protection, clarity
Energetic associations: sun, fire, third eye & solar plexus chakras
Beauty benefits: stimulates hair growth, improves circulation, tones skin, clarifies scalp

How to Grow Rosemary

Type: Rosmarinus officinalis (woody perennial)
Light: Loves full sun—needs 6–8 hours per day.
Soil: Well-drained, slightly sandy soil. Prefers a neutral to alkaline pH.
Water: Drought-tolerant. Water when the top 2–3 cm of soil is dry. Avoid soggy roots.
Spacing: Space 45–60 cm apart if planting multiple shrubs.

Planting Notes
From seed? Yes, but germination is slow (2–4 weeks) and can be inconsistent.

From cuttings? Yes, and highly recommended.
Take a 10–15 cm cutting from new growth.
Strip lower leaves and dip in rooting hormone.
Place in water or moist potting mix. Roots appear in 2–3 weeks.

Harvesting
Harvest small sprigs regularly to encourage bushiness.
Best harvested in the morning when oils are most potent.
Dry by hanging upside down in bundles for teas and DIY use.

Beauty Recipe: Rosemary Hair Rinse (for Growth & Shine)
Rosemary is renowned for stimulating blood flow to the scalp and supporting hair growth.

Ingredients:
- 2 sprigs fresh rosemary (or 2 tbsp dried)
- 2 cups water

How to Use:
Simmer rosemary in water for 15 minutes.
Let cool and strain into a bottle.
After shampooing, pour the rinse through your hair slowly.
Massage it in and leave for 2–5 minutes. Do not rinse out.

Tip: Keep in the fridge for up to a week and apply 2–3 times per week.

Bonus: Rosemary Facial Steam
Add 1 tbsp dried rosemary + 1 tbsp dried chamomile to a bowl of hot water.
Drape a towel over your head and breathe in the herbal steam for 5–10 minutes.
This cleanses pores and revives tired, congested skin.

Energetic Use
Keep a rosemary sprig under your pillow or hang near the entrance of your home for clarity and protection.
Burn a dried bundle when you're ready to clear out mental fog or release old energy.
Whisper while tending it:
"May I remember who I am, and may I never forget how far I've come."

MINT

THE HERB OF VITALITY, FRESHNESS & EMOTIONAL RESET

Mint

Botanical Name: Mentha spp.
Common Names: Peppermint, Spearmint, Garden Mint
Element: Air
Planetary Ruler: Mercury
Season: Spring to Autumn
Energy: Invigorating, Uplifting, Cooling
Companion Plants: Cabbage, Tomatoes, Peas
Beneficial Neighbors: Repels aphids, ants, and flea beetles
Avoid Planting With: Other mints or invasive species (can spread aggressively)
Indoor or Outdoor: Excellent indoors in containers; outdoors with containment.

Spiritual & Cultural Roots

Mint has traveled through time and continents as a symbol of hospitality, healing, and renewal. In Ancient Greece, mint was rubbed on tables to welcome guests. In Egypt, it was prized for its cooling, medicinal properties. In folk magic, mint was carried to lift spirits, draw in prosperity, and banish negativity. Wherever it's grown, mint reminds us of new beginnings—fresh breath, fresh start, fresh mindset.

Magical & Healing Properties

Spiritually used for: energy clearing, upliftment, wealth, travel, clarity
Energetic associations: air, Mercury, heart & throat chakras
Beauty benefits: cools inflammation, brightens skin, soothes itchy scalps, awakens tired muscles

How to Grow Mint

Type: Mentha spp. (perennial herb)
Light: Prefers full sun to partial shade
Soil: Moist, well-drained, fertile soil
Water: Keep consistently moist, but not soggy
Spacing: Grows vigorously—plant 30–45 cm apart or in containers
Important: Mint can be invasive! It will take over garden beds if not contained. Best in pots or planters unless you want a mint jungle!

Planting Notes
From seed? Yes, but can be slow and less consistent
From cuttings? Absolutely! Easy and reliable.
Snip a 10–15 cm stem
Place in water or directly into moist soil

Roots in 1–2 weeks and grows fast

Harvesting
Begin light harvesting once the plant is about 15 cm tall
Snip leaves or stems often to keep plant bushy and vibrant
Best harvested in the morning when essential oils are strongest

Beauty Recipe: Cooling Mint Foot Soak
Perfect for sore feet, swollen ankles, or post-long-day recovery.
Ingredients:
- 1 handful fresh mint (or 2 tbsp dried)
- 1 tbsp Epsom salt
- 1 tsp baking soda
- Optional: 3 drops peppermint essential oil

How to Use:
Steep mint in 2 cups boiling water for 15 mins.
Pour into basin of warm water and stir in salts and baking soda
Soak feet for 15–20 minutes.
Bonus: Rub a mint leaf between your fingers and inhale for a refreshing hit

Bonus: Mint & Aloe Cooling Skin Gel
Soothes sunburn, bites, and overheated skin.
Blend together:
- 2 tbsp fresh aloe vera gel
- 1 tbsp mint-infused water (steep fresh leaves, then cool)

Store in a small jar in the fridge
Apply as needed for relief & radiance

Energetic Use
Burn dried mint in a safe dish to clear stagnant energy or prep your space for a fresh start.
Tuck fresh mint into your wallet or purse for a prosperity boost.
Sip mint tea while journaling to invite truthful self-expression.
Whisper while you water it:
"Let this leaf refresh my soul, and every new day feel brand new

LEMON BALM

THE HERB OF JOY, CALM & INNER LIGHT

Lemon Balm

Botanical Name: Melissa officinalis
Common Names: Balm, Sweet Balm, Bee Herb
Element: Water
Planetary Ruler: Moon
Season: Spring through late Summer
Energy: Calming, Restorative, Uplifting
Companion Plants: Broccoli, Cauliflower, Tomatoes
Beneficial Neighbors: Attracts bees and beneficial insects
Avoid Planting With: Itself (can become invasive if unmanaged)
Indoor or Outdoor: Grows well indoors in pots; prefers full sun outdoors.

Spiritual & Cultural Roots

Known as Melissa officinalis, Lemon Balm has long been the herb of the heart and soul. In Ancient Greece, it was planted around beehives to attract and soothe bees, seen as sacred messengers. In medieval times, monks used it to calm anxiety and sharpen the mind.

Associated with goddesses of love, healing, and the moon, Lemon Balm has always been a gentle ally—bringing comfort in grief, clarity in chaos, and restoration to the weary spirit.

Magical & Healing Properties

Spiritually used for: joy, calm, love, emotional healing, clarity
Energetic associations: moon, water element, heart chakra
Beauty benefits: tones skin, soothes irritation, eases puffiness, supports sensitive skin

How to Grow Lemon Balm

Type: Perennial herb (comes back yearly!)
Light: Full sun to part shade
Soil: Moist, well-drained, rich in organic matter
Water: Water regularly—don't let it dry out completely
Spacing: 30–45 cm between plants
It grows well in garden beds or containers, making it perfect for beginners or balcony gardeners!

Planting Notes
From seed? Yes, but slow to germinate—be patient.
From cuttings? Yes—very easy!
Take a soft stem cutting

Place in water or moist soil
Roots within 1–2 weeks
Transplant into garden or pot

Tip: Keep trimmed to prevent it from going leggy and flowering too early (which can reduce flavour).

Harvesting
Pick leaves regularly once plant is about 15 cm tall
Use fresh or dry by hanging in bunches upside down in a dry, dark area
Best picked in the morning when the oils are strongest

Beauty Recipe: Lemon Balm Facial Steam (for Peace & Glow)
Perfect before bed or during self-care rituals.
Ingredients:
- 1 handful fresh or dried lemon balm
- Optional: 1 tsp dried chamomile or lavender

How to Use:
Boil 4 cups water, pour into a heat-safe bowl
Add herbs, drape towel over your head and bowl
Steam face for 5–7 mins—breathe deeply, relax, let your pores open
Leaves skin soft, clears the mind, and preps you for rest.

Bonus: Joy Tonic for Stressful Days
A beautiful herbal tea blend for emotional support.
Ingredients:
- 1 tsp dried lemon balm
- 1 tsp dried rose petals
- ½ tsp dried peppermint
- Honey to taste

How to Use:
Steep in boiling water 5–7 mins. Sip slowly with intention.
Say: "I drink in peace. I exhale joy."

Energetic Use
Lemon Balm is gentle magic. Burn a dried leaf during a full moon for emotional healing.
Carry fresh leaves in a pouch near your heart for grief support or love energy.
Rub leaves between your hands before journaling to activate clarity and calm.

SAGE

THE HERB OF JOY, CALM & INNER LIGHT

Sage
Botanical Name: Salvia officinalis
Common Names: Garden Sage, Common Sage
Element: Air
Planetary Ruler: Jupiter
Season: Late Spring to early Autumn
Energy: Clearing, Purifying, Strengthening
Companion Plants: Rosemary, Cabbage, Carrots
Beneficial Neighbors: Attracts pollinators and repels cabbage moth
Avoid Planting With: Cucumbers (can inhibit growth)
Indoor or Outdoor: Outdoor preferred; needs strong light indoors.

Spiritual & Cultural Roots
Known as Salvia officinalis, Sage has been used for thousands of years across cultures—from the sacred smudging rituals of Native American nations, to the herbal wisdom of European healers. The name Salvia comes from the Latin salvare, meaning "to heal" or "to save."
In Ancient Rome, Sage was harvested with ceremonial reverence. In Celtic tradition, it was burned for purification and insight. Sage is the herb of wise women, healers, and seers—called upon when clarity is needed and when old energy must be released.

Magical & Healing Properties
Spiritually used for: cleansing, truth, protection, clarity, ancestral connection
Energetic associations: fire + air elements, throat chakra
Beauty benefits: antibacterial, balances oily skin, scalp clarifier, reduces blemishes

How to Grow Sage
Type: Perennial herb (evergreen in some climates)
Light: Full sun—Sage loves the light
Soil: Sandy, well-drained, slightly alkaline
Water: Drought-tolerant once established; don't overwater
Spacing: 45–60 cm between plants
Sage thrives on neglect—it's a resilient garden warrior.

Planting Notes
From seed? Can be tricky—slow to germinate
From cuttings? Yes!
Snip 10–15 cm of new soft growth
Strip lower leaves and place in moist soil or water
Roots in 2–3 weeks

Tip: Prune in spring to encourage bushy growth. Don't let it flower too early—flowers can weaken leaf flavour.

Harvesting
Harvest in the morning for strongest oils
Use fresh or hang to dry in a warm, dark spot
Store in airtight glass jars—use for tea, beauty, rituals

Beauty Recipe: Sage & Apple Cider Scalp Rinse (for Clarity & Strength)
Perfect for oily scalp, dandruff, or spiritual resets.
Ingredients:
- 1 tbsp dried sage
- 1 tbsp rosemary (optional for extra power)
- 2 cups boiling water
- 2 tbsp apple cider vinegar

How to Use:
Steep herbs in boiling water 20 minutes
Strain, let cool, add vinegar
Pour over scalp after shampoo, massage in, rinse or leave in
Promotes healthy scalp, reduces buildup, and clears stagnant energy.

Sacred Smoke: The Ritual of Release
Sage bundles (ethically grown) are burned to purify energy—of people, spaces, and tools.
Ritual Tip:
Light a dried bundle
Fan the smoke gently
Say: "I release what no longer serves me. I invite truth and clarity."
Best used during the new moon, before big decisions, or to cleanse after emotional heaviness.

Clarity Tea Blend: For Focus & Inner Voice
A herbal tea to awaken the throat chakra and sharpen intuition.
Ingredients:
- 1 tsp dried sage
- ½ tsp lemon verbena or mint
- Honey + a squeeze of lemon (optional)

How to Use:
Steep 5 minutes. Drink while journaling, writing, or meditating.
Say: "I speak truth. I see clearly. I trust my wisdom."

Thyme

Botanical Name: Thymus vulgaris
Common Names: Common Thyme, Garden Thyme
Element: Air
Planetary Ruler: Venus
Season: Spring to early Autumn
Energy: Antibacterial, Energizing, Grounding
Companion Plants: Cabbage, Strawberries, Tomatoes
Beneficial Neighbors: Deters pests, especially cabbage worms
Avoid Planting With: Not many known conflicts
Indoor or Outdoor: Can be grown indoors in sunny windows; thrives outdoors in sun.

Spiritual & Cultural Roots

There's an ancient strength in thyme—subtle, aromatic, and steady. It was once strewn across castle floors to purify the air and worn by warriors beneath their armor to give them courage before battle. In ancient Greece, thyme symbolized elegance and bravery, and in medieval times, it was embroidered into garments to inspire strength in the face of fear.

Thyme is a plant of memory and resilience. It thrives in the cracks of stone walls, basking in the sun, holding on even when the soil is thin and the winds are fierce. It teaches us that softness and strength can live side by side—that fragrance can come from struggle, and healing can grow in hard places.

To grow thyme is to embrace quiet tenacity. To use thyme is to invoke a long lineage of medicine women, herbalists, and home healers who knew its magic well. Whether tucked into a bath to ease aching muscles, brewed into a tea for colds and coughs, or infused in oil for its antimicrobial powers—thyme is your steady garden ally, especially when life feels overwhelming.

Growing Thyme

Light: Full sun—thyme thrives best when it has plenty of warmth and light.
Soil: Well-draining, sandy or loamy soil. It dislikes overly wet roots.
Watering: Deep and infrequent—thyme is drought-tolerant once established.
Spacing: Plant 20–30 cm apart. It forms low-growing, woody clumps that spread gracefully.

How to Grow Thyme (Even With Little Experience)

From Seed: Thyme seeds are slow and tiny. Start them indoors on the surface of moist soil. Keep warm and give light. Germination takes 2–4 weeks.

From Cuttings (Best Way): Take soft tips from healthy thyme, remove bottom leaves, and root in water or moist soil. Great for beginners.
From Division: Mature thyme plants can be split and replanted every few years.

Harvest when it begins to bloom, in the morning after dew dries, for the strongest essential oils.

Magical & Cultural Uses
Used for protection, courage, and purification.
Add thyme to your pillow sachets for restful sleep and dreams of clarity.
Burn dried thyme as incense to cleanse a space or banish self-doubt.
Associated with faerie magic—in folklore, thyme helped one see and speak with the fae.

Beauty & Ritual Wellness Uses
Thyme Steam Facial Great for congested, acne-prone skin.
You'll Need:
- A handful of fresh thyme
- A bowl of steaming water
- A towel

How To: Add thyme to the bowl, lean over it with a towel draped over your head, and breathe deeply for 5–10 minutes. Clears pores, calms skin, and clears the mind.

Scalp & Hair Tonic
Thyme boosts circulation, promotes hair growth, and helps with dandruff.
DIY Hair Rinse Recipe:
- 1 cup water
- 1 tbsp dried thyme
- Optional: 1 tsp apple cider vinegar

Simmer the thyme in water, cool, strain, and use as a final hair rinse after washing.

Antibacterial Skin Toner
Make a thyme infusion and apply with a cotton pad to soothe inflammation and fight acne.
Internal Use (With Caution & Guidance)
Thyme tea is used for coughs, sore throats, and digestion.
Acts as a natural antiseptic thanks to thymol.
Helps reduce bloating and supports the lungs.

Always consult with a healthcare provider or herbalist if pregnant, breastfeeding, or taking medication.

Ritual: The Courage Bath
Thyme was always a herb of warriors—let it now be one for your soul.
You'll Need:
- A handful of dried thyme
- A pinch of rosemary
- A few drops of essential oil (lavender or cedarwood)
- A muslin bag or tied cloth

How To: Place herbs into the bag and hang it under the hot water as your bath fills. Add the oil to the tub. Soak in silence. Visualise the water pulling out fear, soaking in steadiness.
Whisper:
"I do not shrink. I do not hide. I face the world with strength and clarity."

Thyme is not just a plant—it's a protector, a cleanser, a quiet reminder that strength can be fragrant, ancient, and enduring.

YARROW

THE HEALER'S BLADE & THE WARRIOR'S SHIELD

Yarrow

Botanical Name: Achillea millefolium
Common Names: Yarrow, Soldier's Woundwort
Element: Water
Planetary Ruler: Venus
Season: Summer
Energy: Balancing, Healing, Protective
Companion Plants: Lavender, Chamomile, Onions
Beneficial Neighbors: Attracts beneficial insects, improves soil
Avoid Planting With: None significant
Indoor or Outdoor: Best outdoors; needs room and airflow to thrive.

Spiritual & Cultural Roots

Yarrow grows like it remembers something. With its delicate white or pink flowers and fern-like leaves, it looks gentle—but beneath that beauty is grit. It's one of the oldest medicinal plants in human history, found in ancient burial sites, battlefield journals, and the garden beds of wise women who knew how to stop blood with a leaf and summon dreams with a tea.

In Greek mythology, Achilles used yarrow to tend to the wounds of his warriors, giving the plant its Latin name Achillea millefolium—a thousand leaves of healing. Across Europe, yarrow was strewn at wedding ceremonies to ensure lasting love, carried into battle for courage, and used in rituals to protect and cleanse sacred space.

Yarrow is the bridge between the healer and the fighter—the soft bloom and the steel within. When you grow yarrow, you grow medicine, both for the body and the soul.

Growing Yarrow

Light: Full sun. Yarrow loves warmth and open skies.
Soil: Well-draining, poor to average soil. It actually prefers less-rich soil for stronger stems and more potent oils.
Watering: Minimal once established—drought-tolerant and very low maintenance.
Spacing: Plant 30–45 cm apart. It spreads easily.

How to Grow Yarrow (Simple, Resilient, Beginner-Friendly)

From Seed: Sow seeds directly outdoors in spring or start indoors 6–8 weeks before last frost. Needs light to germinate—press seeds onto the surface without covering.
From Cuttings: Take softwood cuttings in spring and root them in damp soil.

From Division: Easily propagated by splitting clumps in early spring or autumn.

Yarrow grows strong, spreads quickly, and brings bees and beneficial insects into your garden. A perfect companion for other healing herbs.

Magical & Cultural Uses
Burned to protect against negativity and illness.
Hung over cradles or doorways for spiritual shielding.
Used in divination rituals—yarrow stalks are still used in traditional I Ching readings.
Associated with the heart chakra and the warrior archetype—fierce but compassionate.

Beauty & Wellness Rituals
Yarrow & Rose Healing Facial Steam is perfect for sensitive, inflamed, or acne-prone skin.
You'll Need:
- 1 tbsp dried yarrow
- 1 tbsp dried rose petals
- Bowl of hot water

Place herbs in the bowl, drape a towel over your head, and steam your face for 10 minutes. Yarrow tones and tightens skin while calming redness and fighting bacteria.

Yarrow Wound Healing Salve
Yarrow is legendary for its ability to stop bleeding and speed wound repair.
Salve Recipe:
- 1 cup yarrow-infused oil (olive or sweet almond)
- 2 tbsp beeswax
- Optional: 5 drops lavender essential oil

Melt beeswax into the strained infused oil, stir, pour into jars, and let set. Keep in your herbal first-aid kit.

Scalp Clarifying Rinse
Yarrow tightens pores, calms inflammation, and balances oil production.
- Brew a strong yarrow tea.
- Cool, strain, and use as a final rinse after shampooing.
- Massage gently into the scalp—helps with itchiness, oiliness, and flakiness.

Internal Uses (Traditionally Used With Caution)
Used for colds, fevers, and digestive upset.
Helps regulate menstrual cycles and ease cramps.
Can be made into a tea, tincture, or extract—though not recommended for use during pregnancy.
Always consult a qualified herbalist or healthcare provider when using internally.

Ritual: Warrior's Shield Tea
For strength during emotional or energetic battles.
You'll Need:
- 1 tsp dried yarrow
- 1 tsp lemon balm
- A dash of cinnamon
- A piece of smoky quartz (optional, placed nearby—not in the tea)

Steep herbs in hot water for 5–7 minutes. Sip slowly, visualizing a glowing shield around your heart and body.
Whisper:
"I am held. I am protected. I do not fear the world—I walk through it steady and whole."

Yarrow reminds us that softness is not weakness. It's the choice to bloom, even on broken ground. To stand tall, even when you've been cut down before. To heal, and then help others do the same.

Comfrey

Botanical Name: Symphytum officinale
Common Names: Knitbone, Boneset, Healing Herb
Element: Earth
Planetary Ruler: Saturn
Season: Spring to early Summer
Energy: Restorative, Rooted, Deep Healing
Companion Plants: Fruit trees, Asparagus
Beneficial Neighbors: Builds soil; excellent for compost
Avoid Planting With: Shallow-rooted plants (comfrey dominates space)
Indoor or Outdoor: Outdoor only; deep roots need space.

Spiritual & Cultural Roots

Comfrey is the plant that doesn't just heal—she rebuilds. With her large, deep green leaves and purple bell-shaped flowers, she grows like she means it: big, bold, unstoppable. In herbal medicine, she's known as the "knitbone" herb, revered for her ability to mend bones, soothe inflammation, and regenerate tissue. Her roots run deep, and so does her wisdom.

This is a plant that teaches resilience. That healing is possible even after fracture. That what is broken can grow strong again—if given the right nourishment.

In folk medicine from Europe to Asia, comfrey was used in poultices for sprains and bruises, brewed for internal healing (although modern herbalists now caution against internal use), and always treated with deep reverence. Farmers planted comfrey to feed their soil, nourish their livestock, and offer blessings of abundance.

Comfrey is a mothering plant—rich, generous, grounding. When you grow her, you grow a garden of recovery, renewal, and deep-rooted strength.

Growing Comfrey

Light: Full sun to part shade.
Soil: Rich, moist, well-drained. Thrives in compost-heavy beds.
Watering: Loves moisture. Keep well-watered, especially in dry seasons.
Spacing: Needs room—can grow up to 90 cm tall and wide.

How to Grow Comfrey (From Root, Not Seed)

Comfrey does not grow well from seed and is best propagated via root cuttings or crown divisions.

From Root Cutting: Purchase comfrey root pieces (ideally Bocking 14 variety, which is non-invasive). Plant directly into the soil in early spring or autumn. Bury horizontally, about 5–10 cm deep.

From Division: If you know someone with comfrey, ask for a chunk of their crown in early spring. It transplants beautifully.

Note: Comfrey spreads easily—plant in a permanent spot or use a root barrier. She'll happily take over if you let her.

Magical & Cultural Uses

Traditionally planted around homesteads for health, prosperity, and protection.
Associated with Saturn and earth energy—grounding, slow growth, deep transformation.
Used in spellwork to bring things back together—relationships, emotions, even finances.

Beauty & Wellness Rituals Comfrey Healing Salve (for Bruises, Aches & Skin Recovery)

Ingredients:
- 1 cup comfrey-infused oil (infuse dried or fresh leaves in olive oil over 2–4 weeks)
- 2 tbsp beeswax
- Optional: a few drops of tea tree or lavender oil

Melt beeswax into strained infused oil, pour into tins, and let it set. Use on sprains, sore joints, bruises, and cracked skin.

Comfrey & Rose Recovery Bath Soak

A deeply nourishing ritual for tired bodies and overworked muscles.
You'll Need:
- 1 cup dried comfrey leaves
- ½ cup dried rose petals
- 1 cup Epsom salts

Place herbs in a muslin bag or loose in the bath. Soak for 20 minutes. Let your body receive.

Comfrey Leaf Hair Rinse (for Damaged Hair)

Comfrey is rich in allantoin, which strengthens keratin and promotes hair repair.
- Brew a strong tea with fresh or dried comfrey leaves.
- Cool, strain, and use as a rinse after shampooing.
- Gently massage into scalp and ends—leave for a few minutes before rinsing out.

Great for dry, brittle, or chemically treated hair.

Ritual: The Healing Ground Meditation
Sit by your comfrey plant, or hold a comfrey leaf in your palm.
Breathe in slowly. Feel the weight of your body supported by the earth.
Whisper:
"I am held. I am healing. What was broken is becoming strong."
Visualize your roots sinking deep. The wounds knitting. The earth beneath you wrapping around your spirit like a balm.

Nettle

Botanical Name: Urtica dioica
Common Names: Stinging Nettle, Nettle
Element: Fire
Planetary Ruler: Mars
Season: Spring
Energy: Revitalizing, Nourishing, Detoxifying
Companion Plants: Tomatoes, Mint
Beneficial Neighbors: Attracts beneficial insects, improves compost
Avoid Planting With: Delicate low-growing plants
Indoor or Outdoor: Outdoor only due to its spreading and stinging properties.

Don't let her sting fool you—Nettle is one of the most powerful and nourishing plants in the natural world. She doesn't ask for attention with petals or fragrance. Instead, she stands bold and bristled, demanding respect.

Spiritual & Cultural Roots

This is the plant of wild women. Of boundary-makers. Of strength hidden beneath softness. For centuries, nettle has nourished the blood, soothed inflammation, and supported the body through transitions—from puberty to motherhood to menopause.

She is the mother, the warrior, the protector.

In ancient Europe, nettle was woven into cloth as strong as armor. In Indigenous cultures, she was brewed for blood building and strength. Wherever she grows—quiet, unassuming, on the edge of the road—she brings vitality to the soil and strength to the soul.

Growing Nettle

Light: Partial sun to full shade.
Soil: Moist, rich in nitrogen. She loves compost.
Watering: Consistent watering in dry spells.
Spacing: Give her a corner to herself—nettle spreads.
Caution: Wear gloves when harvesting. Once dried or cooked, the sting disappears, leaving behind her healing essence.

How to Grow

From Seed: Can be started indoors in late winter or directly sown in spring. Seeds need light to germinate, so press gently into moist soil.
From Root Division: The easiest way—cut from an established patch and transplant.

Nettle is perennial and vigorous. Best grown in containers or dedicated spaces if you want to keep her in check.

Magical & Cultural Uses
Protection: Hung in doorways or tucked into clothing to ward off harm.
Fertility & Fire: Associated with Mars and fire element. Sacred to many European folk traditions.
Rebirth & Boundary Work: Used in spells for banishing, strength, and sovereignty.

Beauty & Wellness Rituals
Nettle Hair Growth Rinse (Stimulates Scalp, Strengthens Hair)
You'll Need:
- 2 tbsp dried nettle
- 1 tbsp dried rosemary
- 2 cups boiling water

Steep herbs for 15–30 minutes. Cool, strain, and pour over freshly washed hair. Leave in or rinse lightly. Use 1–2 times weekly.
Nettle is rich in iron, silica, and magnesium—powerful allies for hair growth and scalp health.

Iron-Rich Nettle Nourishment Tea
Perfect for fatigue, menstrual support, and adrenal recovery.
- 1 tbsp dried nettle
- 1 tsp dried peppermint or rose
- 1.5 cups boiling water

Steep 10–15 minutes. Strain and sip. Add honey if desired. Nettle is a true tonic—gentle, but powerful when taken regularly.

Nettle & Green Clay Face Mask (For Purifying and Rejuvenating Skin)
Ingredients:
- 1 tsp nettle powder or steeped tea
- 1 tbsp green clay
- 1 tsp honey or aloe gel

Mix into a paste and apply to clean skin. Leave for 10 minutes and rinse. Leaves skin clarified, toned, and subtly glowing.

Ritual: The Thorn and the Nectar
Sit with nettle. Gloved if need be.
Feel her energy: sharp, wild, unapologetic. This is not a plant that pretends to be

gentle. She stings to remind you of your boundaries.
Say aloud:
"I honour the parts of me that once felt like too much.
I honour my wildness, my wisdom, my fire.
I do not shrink. I do not soften for the comfort of others.
I am Nettle. And I am enough."
Let her remind you that your strength doesn't have to be soft to be sacred.

Nettle is not here to please.
She is here to fortify.
To strengthen your roots.
To sting you awake—into your power.

Cultural & Historical Roots: Plant Lore from Around the World

For as long as humanity has walked the Earth, plants have not only sustained us—they've told our stories.

In every corner of the world, flowers and herbs have been more than just food or decoration. They've been medicine, protection, love spells, offerings, and lineage. They've carried our prayers, marked our ceremonies, and whispered ancestral truths across generations.

In Ancient Egypt, plants weren't merely medicinal—they were divine. Calendula was used in sacred embalming rituals, while aloe was gifted to pharaohs as a plant of immortality. Fragrant oils from herbs like myrrh and frankincense were blended into beauty serums and perfumes for queens like Cleopatra, believed to carry not just scent, but spiritual power.

In Traditional Chinese Medicine (TCM), plants form the core of healing philosophy. Ginseng for vitality. Ginger for digestion. Chrysanthemum for calming the spirit. Each plant is tied not just to symptoms but to energy flow, balance, and seasonal harmony.

In India's Ayurvedic tradition, herbs are soul medicine. Turmeric is sacred, both a culinary staple and a purifying agent in wedding rituals and temple offerings. Hibiscus is crushed into hair oils to bring strength and shine. Neem leaves are hung on doorways to ward off illness and negativity. In Ayurveda, to know a plant is to know its soul.

In Ballardong country, part of the Noongar nation of Western Australia, plant wisdom has been passed down for thousands of years. Native peppermint is used to clear airways, while sandalwood is harvested respectfully for smoking ceremonies and deep healing. Here, nature is never taken—it is listened to, worked with, and honoured. Seasons are marked not by dates, but by changes in wind, animals, and blooms. This is true sustainable living, guided by Country.

In Celtic traditions, herbs like yarrow and mugwort were used by midwives and healers known as wise women or ban-draoí. Rosemary was tucked into baby cribs for protection. Lavender bundles were used in rituals to bless the home. Gardens were places of sanctuary, where plants were companions and teachers.

In Indigenous North and South American cultures, plant knowledge is as sacred as ceremony. White sage and sweetgrass are used to purify, tobacco offered in prayer, and corn, beans, and squash grown together in harmony as the Three Sisters—an interdependent trio of life-giving crops.

And in folk traditions across Europe, you'll find stories of chamomile tea curing sadness, nettle soup building resilience, and thyme laid beneath pillows to guard against nightmares. These weren't simply beliefs. They were truths passed from grandmothers to daughters—truths rooted in lived experience.

Beauty Rituals of the Ancients

Beauty has always been plant-based.

Henna was used in North Africa, the Middle East, and South Asia not just to dye hair and decorate skin, but to celebrate rites of passage and bring good fortune.

Clay masks from riverbeds and earth were used by women across Africa and Indigenous Australia to cleanse the skin and protect against the sun.

In Greece and Rome, rose water and olive oil were used by both men and women for glowing skin and lustrous hair. These weren't luxuries—they were everyday tools for wellbeing.

Even today, many modern beauty products borrow heavily from these ancestral recipes—only now they come bottled and branded. But the truth remains: your garden is your original apothecary.

You Are Part of the Legacy
Whether you're planting mint on a windowsill or creating an herbal scalp rinse from rosemary, you are reclaiming something powerful. You are tapping into thousands of years of human instinct, intuition, and inheritance.

You don't need a formal degree or a grand estate. You only need curiosity, reverence, and a willingness to begin.

Every seed you plant is more than a future flower—it's a thread in a tapestry that has stretched across time and place. It's you, weaving your own story into a shared history of wisdom, care, and wild, untamed beauty.

And in your own rituals—your tea-making, your soil turning, your quiet mornings spent pinching back basil—you are writing the next chapter.

What plants or remedies did your grandmother or family use?
What botanical memories shaped you?
What new traditions are you ready to create?

Why This Matters in Your Garden:
When you plant calendula, it's not just for the balm you'll make. You're stepping into a global circle of healers, midwives, grandmothers, warriors, and mystics.

Planting a garden, then, becomes an act of cultural remembrance and reconnection. It honours not just what grows—but what has always been known.

The Baladong Way

Noongar Wisdom of the Seasons & Earth
Long before gardening apps and seed packets, there was the Earth—and those who listened.
The Ballardong people, Have lived in deep relationship with the land for over 45,000 years. Their connection is rooted in observation, reciprocity, and a profound respect for the natural rhythms of Country.
In Ballardong life, time was never marked by clocks or calendars. It was marked by the flowering of trees, the behavior of animals, the taste of the wind, and the feel of the soil.
And it still is.
Before you plant anything—wherever you are in the world—pause. Listen. Watch. The Earth speaks. And sometimes, she tells a different story than the one we were taught.

Why This Matters Wherever You Are
If you've ever struggled with the traditional "four-season" model of gardening, you're not alone. The Earth doesn't always follow the neat lines of spring, summer, autumn, and winter. In many parts of the world, those seasons don't fully reflect the cycles of the land.
So ask yourself:
- What is my land actually doing?
- What are the birds and animals doing?
- What's blooming right now?
- When does the rain come—and when does it stop?

The answer may not be in a textbook. It may be in your backyard, or better yet —in the wisdom of the First Peoples of your region. In Australia, that means looking to Traditional Custodians who have always known the rhythm of the land.
And in Ballardong Country, those rhythms are reflected in six distinct seasons.

The Six Noongar Seasons

Birak (December–January)
The Season of Fire
Hot and dry. Time of cool burning to renew the land. Red-flowering gums bloom. Seeds are gathered.
Garden Tip: Clear old growth. Plant sun-lovers like basil and coriander.

Bunuru (February–March)
The Second Summer
The hottest time of year. The people move toward water for food and refuge.
Garden Tip: Mulch deeply. Hardy herbs like rosemary and sage thrive.

Djeran (April–May)
Autumn and Early Rain
The land cools. Native wattles bloom. A time of harvest and preparation.
Garden Tip: Great for sowing greens, calendula, parsley, and hardy herbs.

Makuru (June–July)
The Coldest Season
Rain replenishes the land. Emus and kangaroos are hunted. The land rests.
Garden Tip: Focus on composting and nourishing the soil. Collect rainwater.

Djilba (August–September)
Time of First Blossoms
New life begins. Birds breed, and flowers bloom.
Garden Tip: Start planting bee-friendly herbs like chamomile and lavender.

Kambarang (October–November)
The Blooming Season
Wildflowers dominate. Reptiles stir. Warmth returns.
Garden Tip: Feed the soil and begin sowing for summer crops.

Listening to Country
To garden on this land—or any land—is to enter into relationship with it.
You are not the master of the soil. You are its student.
Watch the wind. Notice the ants. Smell the air before a storm. These are not just poetic moments—they are wisdom. They are the Earth teaching you how to live with her, not on top of her.

Planting with Respect
If you are on Noongar land—or any Indigenous land—acknowledge the knowledge that came long before you. Learn what plants are native. Listen to stories if they are shared. Ask how you can grow in a way that gives back more than it takes.
Because when you plant with intention and humility, your garden becomes more than a patch of soil. It becomes a place of reconnection, reverence, and renewal.
You are not just growing herbs.
You are growing legacy.

Sustainability & Stewardship

Composting, Water Conservation & Biodiversity
To garden is not just to grow. It is to give back.
It is a quiet promise whispered to the Earth: I will care for you, as you have cared for me.
Composting: Turning Waste into Life
Composting is magic you can make with your hands. The apple core, the tea leaves, the faded petals—all of it becomes black gold when given time, warmth, and love.
There are many ways to compost—worm farms, bokashi buckets, classic backyard bins. None of them are complicated. What they need most is consistency and care.
- Balance "greens" (veggie scraps, coffee grounds) with "browns" (leaves, cardboard, straw).
- Keep it moist—like a wrung-out sponge.
- Turn it once a week to breathe life into it.

In time, you'll have soil that's rich, earthy, and alive with possibility.

Water Conservation: Every Drop Matters
Whether you're in the red dust of WA or the damp hills of Tasmania, water is life. Let your garden reflect this truth.
- Mulch deeply with straw, leaves, or compost to lock in moisture.
- Choose native or Mediterranean herbs like rosemary, lavender, thyme, and

sage.
- Use watering cans, drip irrigation, or greywater systems to reduce waste.

Biodiversity: A Living Garden is a Thriving Garden
A healthy garden hums with life. It welcomes bees and butterflies, wrens and ladybugs. You don't need to force balance. You just need to invite it.
- Plant flowers like calendula, nasturtium, echinacea, and yarrow.
- Let logs and rocks stay where they fall—they're homes, not clutter.
- Avoid harsh chemicals. The Earth has natural remedies.

Your garden, no matter how humble, is part of a greater whole.
Treat it like sacred ground. Walk barefoot when you can. Speak to your plants. Compost your grief and your coffee grounds. This is stewardship. And it is beautiful.

Companion Planting & Garden Planning

Creating an Eco-Friendly, Harmonious Garden
Nature doesn't monocrop.
She mixes. She layers. She lets things sprawl, intertwine, protect, and even sacrifice themselves for the greater good.
So why should your garden be any different?
Companion planting is an ancient art.
It's about planting not just for beauty or harvest, but for relationships. It's about letting the plants speak to each other—nurture, defend, and thrive together.
This is gardening not as control, but as collaboration.

The Garden Friends: Plants That Help Each Other
Basil + Tomatoes
A classic pair. Basil improves tomato flavour, repels flies and aphids, and creates a little aromatic protection field.
Think of them as the lovers of the veggie world—each better with the other.

Carrots + Leeks
Their strong scents confuse each other's pests. The leek deters the carrot fly. The carrot discourages leek moth.
A quiet but powerful partnership.

Nasturtiums
Beautiful, bold, and brave. They lure aphids away from your veggies and bring in pollinators.
The party guest who distracts all the troublemakers.

Marigolds
Their roots secrete a natural compound that repels nematodes and other pests, while their blooms attract bees and butterflies.
Every garden needs a marigold—they're your gentle warrior.

Lettuce + Chives or Garlic
Chives help deter aphids while garlic keeps away slugs and snails. The lettuce benefits from their presence and grows more peacefully.
Your leafy greens' bodyguards.

Pairs to Avoid: The Not-So-Friendly Neighbours
Onions + Beans
They compete for nutrients and stunt each other's growth. They just don't vibe.

Fennel + Everyone
Fennel secretes compounds that inhibit nearby plants. It's best left to its own corner.

Potatoes + Tomatoes
Both are nightshades and attract the same diseases—planting them together doubles the risk.

Easy Family Groupings to Remember
Think of your plants like families—they often thrive with their cousins:
The Aromatic Allies
Basil, mint, rosemary, thyme, sage—plant near veggies to deter pests.
These are your herbal guardians.

The Legume Lovers
Beans, peas, and lentils enrich the soil with nitrogen. Pair them with hungry greens like spinach or broccoli.
These are your soil healers.

The Leafy Companions
Lettuce, kale, and chard love partial shade—great under taller plants like corn or tomatoes.
These are the understory dreamers.

Joyful Garden Layout Tips
Group by need – Put sun lovers together, shade-lovers in another patch.

Water wisely – Don't mix thirsty plants with drought-tolerant ones.

Mix flowers and herbs throughout – Not just for beauty, but to keep pests guessing and pollinators happy.

Use containers, raised beds, or even repurposed drawers – You don't need a full backyard to grow abundance.

Rotate crops – Avoid disease and soil fatigue by changing what grows where each season.

Plan for joy – Place a bench under wisteria, hang wind chimes in a lemon tree, keep a garden journal nearby.

Let your garden be a living altar. Not just for food, but for presence. Not just for harvest, but for healing.

Sustainability & Stewardship

When & How to Gather With Respect

There is a moment in every gardener's journey when your hands hover above a blooming flower, a trailing vine, a thick-leafed herb—and you wonder, "Is it time?"

Harvesting is more than snipping a sprig of rosemary or collecting mint for tea. It's a conversation, an energetic exchange, a sacred act. To harvest is to say: I see you. I thank you. I carry you forward.

And whether you're growing in raised beds, scattered pots, or a single crate tucked beside your caravan door, the same rule applies: harvest with reverence. This is your guide to knowing when, how, and why to gather in ways that honour the plant, the land, and your own deeper rhythm.

First, Ask Permission

Plants are living beings. They breathe, they feel, they respond to sound, light, and energy. Science is finally catching up to what traditional cultures have always known: that the natural world is aware.

So before you pick, pause.

Place your hand near the plant. Breathe.

Ask silently or aloud:

"May I take some of you today?"

You'll be surprised how often a "yes" feels clear—leaves that lift, a scent that strengthens, a sudden breeze. And when it's a "no," you'll feel that too—tightness, dullness, reluctance in your body. Wait. Listen. Try again another day.

This simple pause rewrites everything. It turns a harvest into a ritual of connection.

How to Harvest: The Gentle Way

No matter your space—be it a sunlit balcony, roadside stop, or backyard wild patch—these universal tips ensure your plants thrive long after the harvest.

Use Clean, Sharp Tools

Always sterilize scissors, shears, or your hands before cutting. This prevents disease spread and shows care. You wouldn't want someone with dirty fingernails digging through your hair, would you? Plants feel the same.

Harvest in the Morning

For most leafy herbs and flowers, early morning is ideal. The oils are most potent, the plant is hydrated from the night, and the sun hasn't yet drained its strength.

Exception: Roots and seeds are often best gathered late in the day or just after rainfall when the soil is soft and the energy has descended.

Take Only a Third

A healthy rule of thumb: never take more than one-third of the plant at once. This ensures it continues to grow and thrive, giving you future harvests and supporting pollinators and the soil ecosystem.

Cut Just Above a Leaf Node

When harvesting stems, cut just above a set of leaves or branching point. This encourages bushier growth and healthier regrowth.

Offer Something in Return

This could be water, a whispered song, a pinch of compost, or your bare feet on the earth. The old ways say: never take without giving back. This exchange keeps the balance whole.

Harvesting in All Living Situations

In a Tiny Home or Backyard:

Use succession planting—sow a few seeds every couple of weeks to keep a rotation going. Group herbs by harvesting needs (daily snips vs. seasonal collections) so you can plan better. Install a small drying rack or use baskets in sunny windows.

In Pots or Balconies:

Pots dry out faster, so always harvest after watering. Use vertical space—hanging baskets or wall-mounted herb planters give you more surface to grow and gather from. Herbs like thyme, mint, and lemon balm thrive in containers and handle frequent harvesting well.

On the Road or in a Caravan:

Portable gardens are powerful. Grow in buckets, hanging shoe organizers, or recycled containers. Harvest often to keep plants compact. Dry herbs by hanging them from curtain rods or laying them in mesh bags on your dashboard (let the sun work for you). Store in small labeled jars with the date and plant name.

Harvesting by Plant Type

Flowers

Best harvested just before they fully bloom—when the petals are vibrant but not fading. Calendula, chamomile, lavender—pluck with fingers or snip with scissors. Lay flat in a single layer to dry in a shaded spot with good airflow.

Leaves (Herbs)

Choose healthy, green, and unblemished leaves. Pick regularly to encourage more growth. If drying, tie in bundles and hang upside-down or place in breathable baskets.

Seeds

Wait until the seed heads are fully dry and start to rattle. Snip and place in paper bags. Label everything! Seeds look surprisingly similar once collected.

Roots

Dig gently with your hands or a trowel. Wash, slice, and dry thoroughly—roots can mould if not properly cured. Store in airtight glass jars in a cool place.

Honouring the Harvest

To honour a plant you've gathered from is to recognize its sacrifice. Even a single leaf contains lifeforce. Here are small rituals that deepen your connection:

- Speak gratitude. A simple "Thank you" changes the energy of the act.
- Return the offcuts. Scatter unused stems, petals, or roots back to the earth to decompose and feed the soil.
- Make a garden altar. Place a stone, feather, or shell near your plants as a sign of care.
- Journal the harvest. Date, moon phase, how you felt. Over time, this becomes your own book of Earth wisdom.

Harvesting as a Mindful Practice

This isn't about being perfect. It's about presence.

Each time you gather from your garden—no matter the size—you're entering into sacred exchange. A reminder that abundance is not just measured in volume, but in relationship.

Whether you're picking thyme from a teacup planter or collecting mint from a wild roadside patch, you're participating in something ancient, something intimate, something holy.

You are not just harvesting plants.

You are harvesting peace, memory, and medicine.

When was the last time you gathered something from the earth with full attention?

What small rituals could you bring into your next harvest to deepen the moment?

Earth-to-Beauty Kitchen

Tools, Storage, Safety & Setup for a Sacred Crafting Space
In every patch of earth, in every tin can of rosemary on a windowsill, there is a potion waiting to be born.
The Earth-to-Beauty Kitchen isn't defined by granite countertops or fancy apothecary cabinets. It's defined by intention. A fold-out bench in a caravan can hold as much alchemy as a polished lab—if the heart behind it beats with care, curiosity, and connection.
Here, we invite you to gather your tools, set up your space, and learn the foundations of turning garden gifts into sacred self-care. Because when you craft your own oils, tonics, and balms—you don't just make products. You reclaim an ancient knowing.

Setting the Scene: Big or Small
Whether you're in a caravan, cottage, or city flat, creating a beauty kitchen is about carving out space—not square metres, but sacred intention.
In a Caravan or Small Space:
- Use collapsible prep tables or trays that fit over your lap or bed.
- Store dried herbs in labelled jars under your seat or in hanging racks.
- Use multi-purpose tools—your coffee pot can double as an infuser, your teacups as mixing bowls.

In a Backyard or Shed:
- Dedicate a corner to your herbal work. A shelf, a crate, or a secondhand table is enough.
- Keep your beauty ingredients separate from cooking ones to avoid cross-contamination.
- Add nature: hang a eucalyptus branch, a small altar, or a grounding stone nearby.

Basic Tools for Your Earth-to-Beauty Kit
You don't need to buy it all at once. Start with what you have, and build slowly. Here's a toolkit you can grow over time:
Essential Utensils:
- Glass jars with lids (for infusions and storage)
- Measuring spoons & cups
- Small stainless-steel saucepan or double boiler
- Mixing bowls (glass or ceramic preferred)

- Sieve or cheesecloth (for straining oils or teas)
- Funnel (makes pouring into small bottles easier)
- Spray bottles, dropper bottles, and tins for final products
- Wooden spoon or spatula (gentle and non-reactive)

Optional Additions:
- Stick blender or mortar & pestle (for creams and pastes)
- Labels and waterproof pens
- A dehydrator or sunny drying rack for herbs

Essential oils (for scent and healing, use sparingly and responsibly)

Safety & Sensitivity in DIY Beauty

Working with nature is powerful. And with power comes responsibility. Here are the golden rules for staying safe and crafting with care:

1. Patch Test Everything

Even natural ingredients can irritate. Before applying anything new to your skin or scalp, dab a small amount on your inner arm and wait 24 hours.

2. Know Your Plants

Not all herbs are safe for everyone. Avoid plants you're allergic to, and research each ingredient's contraindications—especially if you're pregnant, nursing, or on medication.

3. Keep It Clean

Sanitize all tools and containers before use. Wash your hands. Keep pets away during crafting. This ensures your creations last longer and stay safe to use.

4. Label Everything

Date your infusions. Write down recipes and exact ingredients. This isn't just for memory—it's for safety, especially if giving products to others.

5. Use Sterile Storage

Store finished products in amber glass or tins. Keep them in cool, dry, dark places. Avoid plastic where possible—it can leach into your formulations over time.

The Three Sacred Bases: Infuse, Blend, Create

Nearly every DIY beauty ritual begins with one of these steps:

1. Infusions: Water or Oil
   - Herbal Teas & Tonics: Steep fresh or dried herbs in boiling water. Use immediately for rinses or toners, or freeze in ice cubes for future use.
   - Herbal Oils: Place dried herbs in a clean jar, cover with a carrier oil (like olive, almond, or jojoba), and let sit for 2–4 weeks in a sunny window. Shake daily. Strain and store.

2. Blends: Mix & Match

- Combine herbal oils with beeswax or shea butter for balms.
- Stir infused waters with a splash of witch hazel and essential oils for face mists.
- Blend clays, powders, and ground herbs into masks.

3. Create: Ritualize the Process
   - Light a candle or burn some dried rosemary as you begin.
   - Play soft music or work in silence, tuning into the rhythm of the plants.
   - Offer gratitude as you mix—this isn't just crafting. It's communion.

Storage Tips for a Mobile or Minimalist Setup
- Use stackable jars to save space.
- Label lids instead of fronts if they're stored in a box or basket.
- Keep a Beauty Journal: Write down what works, what flopped, and what felt magical.
- Recycle containers: Clean out old face cream jars, spice tins, and dropper bottles to reuse.

If you're on the road, let nature be your assistant—air-dry herbs from your rearview mirror, infuse oils on your dash, and find ingredients as you go. Your beauty kitchen is wherever you are.

Sacred Simplicity: You Don't Need Everything
The old ways never relied on perfection—they relied on resourcefulness.
A single jar of infused calendula oil, stirred with intention, can do more than a shelf full of store-bought serums. A mist made from rosemary tea and love is enough. Let go of pressure. Embrace the power of enough.
This is your Earth-to-Beauty kitchen. May it be humble. May it be holy.

What beauty rituals would you love to reclaim or create?
What tools or spaces can you dedicate to your sacred crafting—even if it's a tray, a basket, or just your hands?

Rituals of Care

Turning Beauty Routines Into Sacred Acts
There's a kind of magic that happens when we slow down. When we wash our hair with intention instead of haste. When we rub oil into our skin like an anointing, not just a step. When we meet our own eyes in the mirror and offer softness, not scrutiny.
This is not about beauty in the commercial sense. It is about care—care as ritual, care as return, care as rebellion in a world that rushes us out of our own bodies.
Rituals of care transform the ordinary into the holy. And whether you live in a

bustling house, a solitary cabin, or a caravan parked beneath gum trees, this sacred practice can belong to you.

Why Ritual Matters
Our ancestors understood this well. They did not separate healing from beauty, or beauty from spirit. To cleanse the skin was to cleanse the soul. To oil the hair was to call in strength. To bathe was to be reborn.
Somewhere along the way, self-care became a checklist. But the Earth has always whispered a slower way. Rituals of care remind us that we are not machines—we are gardens. We bloom with tending, time, and tenderness.
You don't need a spa. You don't need silence. You don't need perfection.
You only need presence.
In a Small Home:
Turn your bathroom or bedroom into a ritual zone. Place a candle beside the sink. Store your beauty blends in woven baskets. Let a plant or shell remind you that you are nature.
In a Caravan:
Designate a small bowl or basket for your beauty items. Light incense or play gentle music before starting. Use mirrors not for judgment, but reflection. Even a basin of warm water becomes sacred when you whisper gratitude into it.
In the Garden or Outdoors:
Let nature be your mirror. Rinse your face under rain or morning dew. Brush your hair under the moonlight. Bathe your hands in a bowl of lavender tea brewed on a camp stove. This is ritual. This is enough.

Daily Rituals: Small Acts with Deep Meaning
These aren't routines to rush through. They are daily returns to your Self.
Morning Oil Ritual
Warm a few drops of infused oil between your palms. Inhale deeply. Gently press into your face, your neck, your heart space. Say something kind to your body. Let this be your armour and your invitation.
Hair Combing Ceremony
Use a wide-toothed comb or your fingers. Begin at the ends and work upward. Imagine brushing out not just tangles—but worry, old thoughts, stress. If using a rinse, pour slowly, like a blessing. Let the water carry away what no longer serves.
Foot Soak for Grounding
In a bowl, combine warm water, sea salt, and a handful of herbs (like mint or rosemary). Soak your feet as you journal, listen to birds, or simply rest. Say: *I root myself. I release. I rise.*

Seasonal & Moon Rituals
Nature lives in rhythm—and so do we. Align your care with the turning of seasons or the lunar cycle to deepen your connection.
New Moon:
A time for renewal. Cleanse your face with a charcoal or clay mask. Journal your intentions. Bathe in rosemary and lemon balm tea.
Full Moon:
A time for abundance and glow. Create a ritual bath with flower petals, epsom salt, and your favourite oil. Anoint your body. Dance, stretch, or simply bask in your own presence.
Seasonal Shifts:
With each equinox or solstice, create a beauty ritual that mirrors the season—cooling mists for summer, warming oils for winter, floral tonics for spring, grounding clays for autumn.

Sacred Practices from Around the World
These global traditions remind us that beauty has always been ritual:
Henna painting (South Asia, Middle East): adorns the body with plant-based blessings.
Smoke cleansing with sandalwood or native herbs (Australia): clears stagnant energy.
Milk baths (Ancient Egypt): soften the skin and soothe the spirit.
Hair oiling with braiding (Africa, India): strengthens both strands and self-worth.
Salt scrubs (Mediterranean, Indigenous Americas): release what is no longer needed.
Each of these is a thread. You can weave them into your life in a way that feels right for you.

How to Craft Your Own Ritual
1. Choose Your Intention: Do you need rest? Courage? Love? Let this guide the ingredients you use and the energy you bring.
2. Select Your Tools: Maybe it's a hair oil, a warm tea, or a spritz of floral mist. Keep it simple.
3. Create Atmosphere: Light a candle. Play a song. Sit in silence. Whatever centres you.
4. Speak to Your Spirit: Say your name. Say your truth. Affirm that this moment belongs to you.

Close with Gratitude: Thank the plants, your body, the space you're in. That gratitude becomes nourishment.

This Is More Than Skin Deep

When you turn beauty into ritual, you begin to remember: you are not here to be fixed. You are not here to perform. You are here to be held. To feel. To soften. To reconnect.

Rituals of care are not indulgent. They are essential. Especially in a world that asks you to forget your body and race through your life.

These acts—slow, gentle, intentional—are how we return to ourselves.

So take your time.

Oil your hair under the stars.

Brew tea in the quiet.

Turn your caravan mirror into a portal of love.

This is your sacred beauty. This is your birthright.

What would your care look like if you treated yourself like sacred land? What could change if every touch was a blessing?

Hair oiling is more than a beauty treatment—it's an ancestral act. Across cultures and centuries, women and men alike have massaged oils into their scalps not just for shine, but for strength. For rest. For ritual. It's a way to ground the mind, feed the follicles, and honour the crown of the body.

This Garden Hair Oil is your invitation to return to this ancient rhythm. Whether you live in a caravan with a compact mirror or in a homestead with a clawfoot bath, this recipe adapts to you. It asks only for care, a few humble tools, and the magic of plants.

What It Does
- Stimulates hair growth
- Soothes dry or itchy scalp
- Strengthens roots and adds shine
- Calms the nervous system through touch
- Replaces synthetic conditioners with pure, earth-based nourishment

Key Ingredients & Their Magic

Rosemary: A stimulant for the scalp. Boosts circulation, strengthens strands, and is known for awakening sluggish follicles.

Lavender: Calming and antimicrobial. Helps relieve itchiness, reduce dandruff, and bring peace to your ritual.

Calendula: Gentle yet powerful. Softens dry scalps, heals minor irritations, and brings sun-kissed energy to your blend.

Carrier Oil (your base):
- Jojoba: Closely mimics the scalp's natural oils. Great for oily or sensitive scalps.
- Olive Oil: Deeply nourishing and widely available. Ideal for dry or thick hair.
- Coconut Oil: Adds shine and reduces protein loss (better in warm climates as it solidifies in the cold).
- Sweet Almond Oil: Light and non-greasy—perfect for fine hair.

What You'll Need
- A clean, dry glass jar with a lid
- Dried herbs: rosemary, lavender, and calendula (fresh herbs can introduce mold unless carefully dried first)
- Your choice of carrier oil
- A fine mesh strainer or cheesecloth
- A dark glass dropper bottle or small jar for storage
- Labels & a permanent marker (for date and blend)

How to Make Your Garden Hair Oil

1. Fill your jar halfway with dried herbs. Equal parts rosemary, lavender, and calendula—or adjust to your personal needs and scent preference.
2. Pour your carrier oil over the herbs until they are completely submerged. Leave about a 2cm space at the top of the jar.
3. Stir gently with a clean, dry spoon to release any air bubbles.
4. Seal the jar and infuse:
   - Sun Method: Place the jar on a sunny windowsill or dashboard (if you're in a caravan or on the road) for 2–4 weeks. Shake daily and whisper a blessing if you like.
   - Heat Method (for quicker results): Use a double boiler to gently warm the herbs in oil for 1–2 hours. Keep heat low to preserve nutrients. Stir occasionally.
5. Strain the oil into a clean glass bowl through cheesecloth or a fine strainer.
6. Transfer to a dark glass bottle for storage. Label with the date and ingredients. Shelf life is about 6–12 months in a cool, dry place.

How to Use

Scalp Massage Ritual
- Warm a teaspoon of oil between your palms.
- Use fingertips to gently massage into your scalp with circular motions.
- Breathe deeply. Take your time. This is your moment.
- Leave on for at least 30 minutes—or overnight under a towel or shower cap for a deep treatment.
- Wash out with a gentle shampoo or shampoo bar. You may need two rounds for thicker oils like coconut.

Hair Tips or Ends Only
- Apply a few drops to dry ends to prevent breakage and split ends. No rinse needed.
- Perfect for frizz control or post-sun protection.

Tips & Variations
- Add a few drops of essential oil (like peppermint or cedarwood) to boost benefits—but go lightly. 3–5 drops per 100ml is plenty.
- If your scalp is very sensitive, do a patch test first. Everyone's skin is unique.
- Keep your oil free from water or fingers to extend shelf life.

Rooted in Ritual

This isn't just a hair oil. It's a potion of remembrance. As your fingers move across your scalp, you're connecting to grandmothers and wise women, to dreamers and healers, to yourself.

Let this oiling be your time to reconnect. To reflect. To rest.
You don't need perfect hair. You need sacred hands and a little oil.
That is enough.

HERBAL SCALP TONIC

REFRESH, REVIVE & REBALANCE

Your scalp is the soil from which your hair grows. And just like any good soil, it needs tending—cleansing, nourishing, and occasional revitalisation.

This Herbal Scalp Tonic is a water-based blend that brings balance back to your roots. Whether you've been sweating in summer heat, dealing with flakes, or feeling energetically heavy, this tonic is your go-to. It's light, fast-absorbing, and perfect for mobile living—especially when you don't have time or space for full hair washing.

Use it as a scalp mist, a rinse, or a compress. It's especially helpful for those in caravans, tiny homes, or off-grid living where water is precious and showers are less frequent.

What It Does
- Clears buildup and refreshes the scalp
- Soothes irritation or itchiness
- Strengthens follicles and stimulates hair growth
- Balances oil production
- Feels like a spa in a bottle (even if you're in your pyjamas by a creek)

Key Ingredients & Their Benefits

Nettle: Rich in minerals like silica and iron. Strengthens hair, reduces shedding, and balances the scalp.

Peppermint: Cooling and stimulating. Increases circulation to the scalp and clears blocked follicles.

Chamomile: Calming and anti-inflammatory. Great for sensitive skin and bringing peace to irritated scalps.

Apple Cider Vinegar (ACV): Gently exfoliates, restores pH balance, and smooths hair cuticles. Don't worry—the vinegar smell fades fast.

What You'll Need
- 1 tablespoon each of dried nettle, chamomile, and peppermint (or fresh if well-rinsed)
- 1 cup boiling water
- 1 tablespoon apple cider vinegar (with "the mother")
- A strainer or cheesecloth
- A spray bottle or small jar
- Optional: 2–3 drops of essential oil (lavender, tea tree, or rosemary)
-

How to Make Herbal Scalp Tonic

1. Infuse the Herbs:

 Place your herbs in a bowl or heat-safe jar. Pour 1 cup of boiling water over them. Cover with a lid or plate to trap the steam and nutrients. Let steep for 30–

60 minutes.

2. Strain Well:
Once cooled, strain the infusion into a clean bowl using cheesecloth or a fine sieve.

3. Add ACV:
Stir in your tablespoon of apple cider vinegar. This step is vital—it helps balance scalp pH and adds a natural shine.

4. Bottle It:
Transfer your tonic to a clean spray bottle or jar. Label with ingredients and date. Store in the fridge for up to 1 week (longer if using preservatives, but we prefer fresh).

How to Use It

As a Scalp Spritz:
- Shake well.
- Spray directly onto the roots of clean or second-day hair.
- Massage gently with fingertips.
- Let it absorb—no rinse required.

As a Leave-In Rinse:
- After shampooing, pour or spray over the scalp and hair.
- Gently towel dry or air dry—no need to rinse unless desired.

As a Soothing Compress:
- Soak a cloth in the tonic.
- Press onto the scalp while lying down or resting.
- Ideal for itchy or overheated scalps.

Tips for Nomadic Living
- No fridge? Keep it cool and use within 2–3 days. Add a drop of vitamin E oil or a splash of witch hazel to extend shelf life slightly.
- Use ice cube trays to freeze leftover tonic—just thaw one cube at a time as needed.
- Keep a small travel-size bottle in your bag for on-the-go scalp love.

This Is More Than Haircare

This tonic isn't just a blend—it's a balm for your nervous system. The act of misting your scalp, pausing, breathing—it's medicine.

And in a world that pushes us to rush, to wash and go, to do more and feel less—this tonic brings you back. Back to your roots. Back to your breath.

It's not just about having "good hair." It's about feeling held.

Some remedies don't need explanation—they just work. Passed down from garden witches and wise grandmothers, the herbal salve is one of the oldest and most beloved tools of natural care.

Whether you call it balm, ointment, or medicine-in-a-tin, this little jar of green-gold carries deep magic. It soothes cracked heels and grazed elbows, calms insect bites, comforts burns, and nourishes dry skin that's weathered sun, wind, or labour.

Made from simple ingredients you can grow or gather, this Healing Salve belongs in every tiny home, caravan cupboard, pocket, and palm. It's rugged and gentle. Earthy and elegant. A true companion for the skin and spirit.

What It Does
- Soothes inflammation, stings, and bites
- Speeds healing of cuts and scrapes
- Softens dry, cracked, or weather-exposed skin
- Acts as a protective barrier while allowing skin to breathe
- Invokes comfort, calm, and ancient earth care

Hero Ingredients & Why They Matter

Calendula: Anti-inflammatory and skin-healing. Known as the "skin's best friend," it encourages repair, reduces redness, and calms irritation.

Comfrey: Accelerates tissue regeneration. Traditional name "knitbone" says it all—helps close small wounds and support bruises.

Plantain Leaf: Nature's bandage. Excellent for bites, stings, splinters, and hot, angry skin.

Lavender: Antiseptic and calming. Adds a gentle floral scent and soothes both the skin and the mood.

Beeswax: Locks in moisture, creates a protective seal, and gives your salve that perfect balm texture.

Carrier Oil (Olive, Almond, or Infused Herbal Oil): Extracts the plant medicine and delivers it deeply into the skin.

What You'll Need
- 1/2 cup dried herbs (a mix of calendula, comfrey, plantain, and lavender works beautifully)
- 1 cup carrier oil (olive, almond, or pre-infused oil)
- 1/4 cup beeswax pellets or grated beeswax
- Optional: 5–10 drops essential oil (lavender or tea tree for extra healing)
- A small saucepan or double boil
- Cheesecloth or fine strainer

- Small tins or glass jars with lids
- Labels and date marker

How to Make It

Step 1: Create Your Herbal Oil
(If you don't already have a pre-infused one)
- Place your dried herbs in a clean jar.
- Cover fully with oil, leaving space at the top.
- Let infuse for 2–4 weeks in a warm place (shake daily).

OR
- Gently heat in a double boiler on low for 1–2 hours to speed up the process. Stir occasionally.

Step 2: Strain Your Oil
- Once infused, strain through cheesecloth or fine mesh into a clean jar. Squeeze every last drop—it's precious!

Step 3: Melt and Mix
- In a small saucepan or heat-safe bowl over a pot of water, combine your strained oil and beeswax.
- Heat gently until fully melted, stirring with a clean utensil.
- Add essential oils now, if using.

Step 4: Pour and Set
- Carefully pour your hot mixture into tins or jars.
- Let cool, uncovered, until firm (about 30 minutes).
- Once cool, label and store in a cool, dry place. Shelf life: 9–12 months.

How to Use It
- Rub gently onto scrapes, dry skin, insect bites, or windburned cheeks.
- Massage into heels or hands after gardening.
- Apply a thin layer over cracked lips or minor burns.
- Use as a spiritual salve, massaging into the chest during anxious moments or before sleep.

Good to Know
- Don't use on deep, open wounds or punctures—this salve is for surface healing only.
- Always patch test first if using on sensitive skin or children.
- If you're vegan, candelilla wax is a plant-based alternative to beeswax.

Ritual Tip
Before sealing your jars, whisper a blessing over your salve. Something like:

"May this balm bring peace, healing, and protection to all who use it."
Your hands carry energy. Let it be kind.

This Is Earth Medicine
This isn't just skincare. It's ancestral wisdom sealed in a jar. It's your garden saying, "I've got you." It's a reminder that healing is close—sometimes growing right outside your door.
So carry this salve like a talisman. Use it with intention. Share it with love.
You are a healer now, too.

A flower mist is more than water. It's dew collected from dreams. It's petals captured in droplets. It's your daily invitation to pause, soften, and breathe.

This Flower-Infused Face Mist is a ritual in a bottle. You can use it as a toner, a midday refresh, or a moment of calm when life feels dusty and fast. It carries the spirit of the garden—cool, fragrant, and full of grace.

Whether you live in the dry heat of a van in WA or near a mossy forest cabin, this mist belongs in your beauty ritual. It's quick to make, easy to store, and deeply beautiful in both function and feel.

What It Does
- Gently tones and hydrates the skin
- Cools and calms the face (especially in hot or dry climates)
- Brightens your mood with the scent of herbs and flowers
- Doubles as a linen or aura spray for energy clearing
- Connects your skincare to the rhythm of plants and presence

Star Ingredients & Their Blessings

Rose Petals: Hydrating, anti-inflammatory, and full of love. Rose soothes the skin and the heart.

Chamomile: Gentle on sensitive skin. Reduces redness, puffiness, and irritation. Brings peace.

Lavender: Antibacterial and calming. Helps balance oil and ease blemishes.

Witch Hazel (alcohol-free): A natural astringent. Tightens pores, tones skin, and extends shelf life.

Distilled Water: A clean base that won't encourage bacteria growth.

What You'll Need
- 1 tablespoon each of dried rose petals, chamomile, and lavender
- 1 cup boiling distilled water
- 1–2 tablespoons alcohol-free witch hazel
- Optional: 3–5 drops essential oil (rose, lavender, or frankincense)
- A small spray bottle (amber or cobalt preferred for light protection)
- Fine mesh strainer or cheesecloth
- Funnel (for easy pouring)
- Label and date marker

How to Make It

1. Steep Your Botanicals

Add your dried flowers to a heat-safe jar or bowl. Pour 1 cup boiling distilled water over them. Cover with a lid or plate and steep for 20–30 minutes to capture their essence.

2. Strain the Infusion

Strain through a fine mesh sieve or cheesecloth into a clean jug or bowl. Compost the spent flowers—they've given their gift.

3. Add Witch Hazel & Oils (Optional)

Stir in the witch hazel. Add a few drops of essential oil if desired for scent or added benefit. Stir with a clean spoon.

4. Bottle & Store

Using a funnel, pour your blend into a clean spray bottle. Label it with the date and ingredients. Store in the fridge and use within 1–2 weeks for freshness. (If you add a natural preservative, it can last longer.)

How to Use
- Spritz over clean skin morning and night as a toner
- Use throughout the day to cool and refresh
- Spray onto your pillow before sleep or meditation
- Keep a travel-size mist in your bag for road trips or hot afternoons
- Mist over your scalp, hair, or wrists for a burst of botanical peace

Storage Tips for All Living Spaces
- If you don't have a fridge, store in the coolest, darkest spot available and use quickly
- For extended shelf life: add a teaspoon of vodka or natural preservative like Leucidal
- Always shake before use and discard if scent or texture changes

More Than a Mist

Each spray of this mist is a moment of reconnection. It says: Pause. Feel your skin. Breathe in the garden. Whether you're in a hurry or a ritual, this little bottle helps you return to presence.

It's not just about toning your skin. It's about toning your spirit. Cooling your worries. Inviting in softness.

So go ahead. Mist like you mean it.

This is your garden's answer to plastic-free haircare: the Root & Petal Shampoo Bar. Compact, zero-waste, and rich in botanical love, this bar transforms herbs and natural clays into a cleansing ritual that respects both your hair and the planet.

No synthetic fragrances. No sulphates. No waste. Just plants, roots, petals, and earth—blended into a bar that travels wherever you go, from caravan showers to rainwater baths under the stars.

What It Does
- Gently cleanses scalp and hair without stripping natural oils
- Supports healthy growth and scalp balance
- Leaves hair soft, strong, and naturally fragrant
- Replaces plastic bottles with beauty that biodegrades
- Perfect for travel, off-grid living, or minimalist lifestyles

Star Ingredients & Their Role

Rhassoul Clay (or Bentonite Clay): Gently draws out impurities while adding minerals that strengthen hair strands. Leaves hair feeling clean but not dry.

Marshmallow Root: Rich in mucilage. Conditions, detangles, and soothes the scalp.

Rosemary: Stimulates the scalp and encourages growth. Adds shine and strength.

Lavender: Calming, antimicrobial, and beautifully aromatic.

Coconut Oil: Adds moisture and a gentle lather. (Use fractionated coconut oil if living in cooler climates where solid oil hardens.)

Shea Butter or Cocoa Butter: Deeply nourishes and protects.

Lye (Sodium Hydroxide): Used in traditional soap making (cold process). Essential for creating a solid bar—but disappears completely during the process.

Note on Lye Safety

Working with lye requires caution. It's caustic in raw form but perfectly safe once saponification is complete. Always wear gloves, goggles, and work in a well-ventilated space. If you'd prefer to skip lye, a melt-and-pour soap base is a beginner-friendly alternative (let me know if you'd like a separate version for that!).

What You'll Need (Cold Process Method)
- 200g coconut oil
- 100g shea butter or cocoa butter
- 100g olive oil
- 60g dried herbs (marshmallow root, rosemary, lavender)

- 1 tbsp Rhassoul or Bentonite clay
- 65g lye (sodium hydroxide)
- 170g distilled water
- Optional: essential oils (20–40 drops total)
- Safety gear: gloves, goggles, apron
- Stainless steel or glass bowl
- Silicone soap mold or lined container
- Stick blender
- Digital scale
- Heat-safe jug or container for lye solution

How to Make It (Cold Process)

1. Infuse the Oils (Optional but lovely):
Warm your oils gently and add dried herbs. Let sit for 1–2 hours on very low heat, then strain. This infuses your base with plant power.
2. Prepare Lye Solution:
   - Put on your safety gear.
   - In a well-ventilated area, slowly add lye to water (never the reverse) while stirring. It will heat up.
   - Set aside to cool to around 38–43°C (100–110°F).
3. Melt Oils:
Gently melt the butters and combine with the infused or plain oils. Let cool to the same temperature range as the lye solution.
4. Combine & Blend:
Slowly pour the lye solution into the oils. Use a stick blender to blend until it reaches "trace" (like a thin custard texture).
5. Add Clay & Essential Oils:
Mix in the clay and any essential oils (lavender, rosemary, tea tree). Stir well.
6. Pour Into Molds:
Pour the mixture into molds. Tap gently to release air bubbles. Cover with a towel and let set for 24–48 hours.
7. Unmold & Cure:
Remove from molds and place bars on a rack. Let cure in a dry, airy space for 4–6 weeks. This allows saponification to fully complete and the bar to harden.

How to Use

- Wet your hair and the bar.
- Rub gently between your hands or directly onto scalp.
- Massage in circular motions. Let the herbs and clay do their work.
- Rinse thoroughly. Follow with a rinse or conditioner bar if desired.

Storage Tips
- Keep bar dry between uses—use a slatted soap dish or pouch.
- Perfect for travel—no liquids, no leaks, no waste.
- Store extras in a cool, dry place, wrapped in wax paper or muslin.

Why This Matters
Every bar you make is one less plastic bottle in the landfill. One more act of self-care rooted in the Earth. One more reminder that simplicity is sacred.
This shampoo bar isn't just practical—it's poetic. It carries herbs, history, and heart.

A Note from the Heart
While traditional shampoo bars are often made using lye (sodium hydroxide) to create a solid soap, I want to be transparent:
I'm not a fan of lye.
It might be natural in the technical sense, but to me—it feels like Drano. It's a heavy, caustic chemical, and the energy of it doesn't align with the soft, sacred purpose of this book.
If you feel the same, you're not alone—and there are better options.
You can create melt-and-pour shampoo bars using natural, pre-saponified soap bases that don't require handling lye. Or explore syndet bars made from gentle plant-derived cleansers.
The goal is simple: clean hair, clean conscience, no compromise.
If you're interested, I'll gladly include a no-lye alternative version that's beginner-friendly and just as beautiful.

GREEN CLAY & HERB FACE MASK

DETOX, SOOTHE & SINK INTO STILLNESS

There is something ancient and pure about smoothing earth onto your skin. Clay, long before it became a trend, was used by cultures all over the world to cleanse, calm, and awaken the face. It holds the memory of mountains, the minerals of time, and the power to draw out what no longer serves.

This Green Clay & Herb Face Mask blends nature's deep detox with gentle herbs to create a ritual that nourishes—not strips. It's perfect for days when your skin feels heavy, dull, or in need of a reset. But more than that, it's a sacred moment to be still, to rest, and to remember you are of the Earth.

What It Does
- Draws out impurities and excess oil
- Soothes redness and inflammation
- Tightens pores and tones the skin
- Gently exfoliates and restores glow
- Feels like a mud bath for your soul

Key Ingredients & Their Alchemy

Green Clay (French or Australian): Detoxifying and mineral-rich. Absorbs oil, draws out toxins, and revitalizes tired skin.

Chamomile: Calming and anti-inflammatory. Ideal for sensitive or reactive skin.

Lavender: Antimicrobial and soothing. Brings both scent and skin support.

Calendula: Gentle healer. Helps with redness, minor irritations, and restoring skin tone.

Rose Petals (optional): Softens skin and opens the heart.

Distilled Water, Herbal Tea, or Floral Hydrosol: The mixing liquid—choose one based on what you have or need. Each changes the energy of the mask.

What You'll Need
- 2 tablespoons green clay
- 1 teaspoon each of dried lavender, chamomile, and calendula (grind or powder if possible)
- Optional: 1 teaspoon rose petals or rose powder
- 2–3 tablespoons water, cooled tea (like chamomile or green tea), or hydrosol (like rose or lavender)
- Small bowl and spoon
- Optional: 1–2 drops of essential oil (lavender or tea tree)
- Clean facecloth and warm water

How to Make It
1. Blend Your Base

In a small bowl, mix your green clay with the powdered herbs. If using whole dried herbs, grind them in a mortar and pestle or blender first for a smoother texture.

2. Add Liquid Slowly

Drizzle in your chosen water, tea, or hydrosol a little at a time while stirring until you reach a thick, spreadable paste. Think: smooth mud, not soup.

3. Add Essential Oils (Optional)

If you want an aromatherapy boost, add 1–2 drops of essential oil. Stir with intention.

4. Use Immediately

This mask is meant to be made fresh. It doesn't store well without preservatives, so make just enough for one use.

How to Use

Start with a clean, damp face.
Apply mask with fingers or a brush, avoiding eyes and lips.
Let sit for 5–10 minutes only—do not let the mask crack or fully dry (this can dry out your skin).
Spritz with water or mist if needed to keep it slightly moist.
Rinse gently with warm water and a soft cloth. Pat dry. Follow with a hydrating oil or face mist.

Clay Wisdom
Clay is powerful—use once a week or when needed, not daily.
Always avoid metal bowls or spoons (clay can react with them). Use wood, glass, or ceramic.
Patch test first if your skin is sensitive or new to clay masks.

Tips for Small Spaces or Caravans
Pre-mix dry ingredients into a small jar. Add liquid only when ready to use.
Use leftover tea from the morning as your mixer—nothing wasted.
Keep a mini travel bowl and spoon in your toiletry kit for on-the-road rituals.

More Than a Mask

When you mix clay and herbs, you're not just making skincare. You're crafting an offering. You're pressing pause. You're letting the Earth hold your face in her cool, mineral-rich palms.

Let each application be a reminder:
You are worth the time.
You are not meant to be rushed.
You are allowed to rest.

MOON BALM FOR RITUAL BATHS

ANOINT, MELT, AND SOAK IN THE SACRED

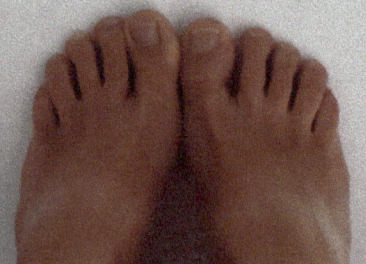

There is something primal, ancient, and transformative about bathing. Not just the act of getting clean, but the act of letting go. Of stepping into water with intention. Of calling in the moon, the earth, and your own inner tide.

This Moon Balm isn't a lotion. It's not quite a butter. It's something more poetic —a meltable balm you can massage into your skin before a bath, or drop into the water like a spell. Rich with oils, herbs, and soft scent, it's your invitation to unwind not just your body, but your spirit.

It's self-care wrapped in ritual. It's a salve for the soul. And whether you soak in a clawfoot tub, a basin under the stars, or a hot bucket in your caravan, this balm brings ceremony wherever you are.

What It Does
- Moisturizes and softens skin
- Infuses your bath with herbs and oils
- Creates a moment of sensual stillness and connection
- Supports lunar rituals, intention setting, and release
- Doubles as a body balm or massage butter outside the bath

Herbal Allies & Lunar Energy

Lavender: Calms the nervous system, quiets the mind, and invites softness.
Chamomile: Soothes skin, relaxes muscles, and gently uplifts.
Rose: Opens the heart. Encourages emotional healing, self-love, and divine feminine energy.
Coconut Oil: Melts easily and nourishes skin deeply.
Shea Butter: Rich and healing. Seals in moisture and softens roughness.
Essential Oils (optional): Frankincense for depth, ylang ylang for sensuality, or clary sage for dreamwork.

What You'll Need
- 1/2 cup coconut oil
- 1/4 cup shea butter
- 2 tablespoons dried herbs (a mix of lavender, chamomile, and rose petals works beautifully)
- 10–15 drops essential oil (optional)
- 1 tablespoon beeswax (for a firmer balm, optional)
- Cheesecloth or strainer
- Heat-safe bowl and saucepan (or double boiler)
- Clean tin or glass jar with lid
- Spoon or small spatula
- A quiet hour, a lit candle, and a whisper of moonlight

How to Make It

1. Infuse the Oils

In a double boiler or heat-safe bowl over simmering water, gently melt the coconut oil and shea butter together. Add dried herbs and let infuse for 30–60 minutes over low heat, stirring occasionally.

2. Strain with Love

Strain the warm oil through cheesecloth into a clean bowl. Squeeze the herbs to get every drop of that golden goodness.

3. Add Beeswax (Optional)

If using beeswax for a firmer balm, return strained oil to the heat and add the beeswax. Stir until melted.

4. Add Essential Oils

Once removed from heat, stir in your essential oils (if using). Breathe them in. Let their scent set your intention.

5. Pour & Set

Pour your mixture into a clean jar or tin. Let cool until solid. Label with a name, date, and maybe even the moon phase you crafted under.

How to Use

- Pre-bath anointing: Rub a small amount over arms, legs, belly, and chest before soaking. Let the oils melt into the water and your skin.
- Drop into bath: Scoop a spoonful into hot water and swirl gently. Inhale deeply. Let the water hold you.
- Massage balm: Use as a body butter post-bath or pre-bed, especially over the heart or womb space.
- Lunar ritual: Use on New Moons for intention-setting or Full Moons for release.

More Than Moisture

This balm isn't just about silky skin. It's about coming back to yourself. About making the mundane magical. About honouring the cycles that move through you—like the moon, like the tide, like the garden.

You don't need a spa. You don't need hours. You just need you and a spoonful of moonlight.

Let this balm be your blessing.

There was a time—before salons, before boxes of chemicals, before foils and fumes—when hair was coloured by the Earth herself. Roots, leaves, flowers, and bark were gathered with intention and brewed into dyes. Hair colouring wasn't just aesthetic—it was ceremonial. It marked rites of passage, sacred unions, seasonal shifts, and inner transformations.

Whether you're looking to deepen your brunette, add golden warmth, soften greys, or simply connect with the magic of natural beauty, earth-dyed hair is your quiet rebellion. No bleach. No peroxide. No harsh smells. Just plants, patience, and reverence.

Why Go Natural?
- Free from synthetic dyes and toxic chemicals
- Safe for sensitive scalps, pregnant women, and the planet
- Blends gracefully with greys instead of covering harshly
- Adds shine, depth, and subtle hues
- Encourages slowness and ritual in beauty

The Ritual of Colouring
Plant-based hair dye isn't instant. It's a journey. The colour deepens over hours —or days. It's like brewing tea for your strands, not painting them.
That's the beauty of it.
It's slow. It's sacred. It's yours.

Recipe: Walnut & Rosemary Brunette Rinse
A soft, earthy way to darken hair and add depth to greys
- Enhances natural brunettes and deepens light brown tones
- Adds luster, soft warmth, and cool undertones to greys
- Encourages shine and scalp health
- Builds colour over time (won't give you instant jet-black)

Ingredients
- 1/2 cup black walnut hulls (dried and crushed—NOT the nut meat)
- 2 tablespoons rosemary leaves (fresh or dried)
- 2 cups water
- Optional: 1 tablespoon sage (for extra grey blending)
- Optional: 1 tablespoon apple cider vinegar (helps set colour and shine)

You'll Also Need
- Small saucepan
- Fine strainer or cheesecloth

- Squeeze bottle, spray bottle, or bowl
- Old towel or t-shirt (to protect from drips or staining)
- Gloves (walnut stains skin!)
- Shower cap or plastic wrap (to help colour absorb)

How to Make It

1. Simmer the Brew

In a small pot, combine walnut hulls, rosemary, and sage (if using). Cover with water and bring to a simmer. Let simmer for 30–45 minutes.

2. Cool & Strain

Remove from heat. Allow to steep as it cools—ideally another 30 minutes. Strain through cheesecloth into a bowl or jug. Stir in apple cider vinegar if using.

3. Bottle or Use Fresh

Transfer to a bottle or use immediately for best potency. This rinse will last 5–7 days in the fridge.

How to Use

For a subtle stain / daily rinse:
- Pour over clean, damp hair.
- Catch excess in a bowl and reapply 2–3 times.
- Let sit for 15–20 minutes. Rinse lightly or leave in.

For deeper colour (especially on greys):
- Apply generously with gloves.
- Cover with a shower cap or plastic wrap.
- Leave on for 1–2 hours.
- Rinse and air dry. Repeat weekly to build tone.

Safety Notes

- Do a patch test 24 hours before using.
- Walnut hulls can cause allergic reactions in people with tree nut sensitivities—do not use if allergic.
- Avoid getting the rinse on surfaces—it will stain.

Alternatives for Other Tones

Golden Highlights: Chamomile tea, lemon juice, and sunshine (best on light hair)
Warm Auburn: Strong hibiscus tea + rooibos
Deep Red: Henna (body art quality, mixed with warm tea)
Ashy Brown: Sage leaf and nettle rinse
Black Shine: Indigo (requires prep and layering with henna)

This Is More Than Colour
When you dye your hair with plants, you enter into a relationship—not a transaction.
You stir the pot. You steep the herbs. You wait. You listen. And then—slowly—you change. Like the seasons. Like the trees.
This isn't about covering who you are. It's about becoming more of you.

Henna Hair Dye (Lawsonia inermis)

Regions of Origin: North Africa, the Middle East, South Asia

Functional Benefits: Natural pigment for reddish tones; provides strengthening and conditioning properties.

Materials:
- 100g henna powder (adjust quantity per hair length)
- Warm water
- 1 tbsp coconut oil (optional for moisture)
- Juice of half a lemon (optional for tonal brightness)

Protocol:
1. Prepare a thick paste by combining henna with warm water.
2. Optional: Add lemon juice for enhanced color and coconut oil for added hydration.
3. Allow the mixture to oxidize and activate for 4 hours.
4. Apply evenly to sections of hair.
5. Cover hair and allow to process for 2–4 hours.
6. Rinse thoroughly using warm water.

Hibiscus Hair Mask (Hibiscus sabdariffa)

Regions of Use: Ayurvedic and tropical pharmacopoeias

Functional Benefits: Enhances follicle strength, reduces hair fall, promotes luster

Materials:
- Fresh hibiscus petals and leaves
- 1 tbsp coconut oil

Protocol:
1. Grind plant material into a uniform paste.
2. Blend in oil for enhanced emollience.
3. Apply to scalp and hair shafts.
4. Leave on for 30 minutes; rinse thoroughly.

Fenugreek Hair Paste (Trigonella foenum-graecum)

Region of Use: India

Functional Benefits: Protein-rich; improves elasticity, reduces breakage

Materials:
- 2 tbsp fenugreek seeds
- Water for soaking
- 1 tsp honey (optional)

Protocol:
1. Soak seeds overnight to soften.
2. Blend into a paste.

4. Integrate honey for added humectant effect.
3. Apply to scalp and hair; leave for 30 minutes.

Shikakai Cleanser (Acacia concinna)
Region of Use: Ayurveda
Functional Benefits: Natural saponin content for cleansing; enhances shine
Materials:
- 2 tbsp shikakai powder
- 1 cup warm water
- 1 tbsp aloe vera gel (optional)

Protocol:
1. Mix ingredients into a smooth paste.
2. Distribute evenly across scalp.
3. Leave for 15–20 minutes.
4. Rinse thoroughly.

Ritha Shampoo (Sapindus mukorossi)
Region of Use: South Asia (Ayurveda)
Functional Benefits: Natural surfactant alternative; gentle and non-stripping
Materials:
- Handful of Ritha berries
- 2 cups water

Protocol:
1. Soak berries overnight.
2. Simmer the mixture for 15 minutes.
3. Strain and utilize the liquid as shampoo.

Brahmi Scalp Oil (Bacopa monnieri)
Region of Use: Ayurveda
Functional Benefits: Supports cognitive function; reduces scalp inflammation
Materials:
- 2 tbsp dried Brahmi
- 1 cup coconut oil

Protocol:
1. Gently warm mixture for 30 minutes to infuse.
2. Strain.
3. Apply to scalp as a nighttime treatment.

Amla Rinse (Phyllanthus emblica)
Region of Use: South Asia
Functional Benefits: Rich in antioxidants; improves shine and scalp health
Materials:
- 2 tbsp amla powder or fresh juice
- 1 cup warm water

Protocol:
1. Mix into solution.
2. Apply post-shampoo.
3. May be rinsed out or left as a leave-in.

Beetroot Tint
Contemporary Usage: Botanical pigment alternative
Functional Benefits: Temporarily enhances red or auburn hues
Protocol:
1. Blend beetroot juice.
2. Mix with henna or conditioner.
3. Apply and leave for 20 minutes.

Coffee Rinse
Ethnobotanical Context: Folk applications for color enhancement
Functional Benefits: Adds depth to brunettes
Protocol:
1. Brew strong black coffee.
2. Allow to cool.
3. Use as a rinse or coloring enhancer.

Chamomile Highlight Mist
Ethnobotanical Context: Traditional lightening method for fair hair tones
Protocol:
1. Brew chamomile tea.
2. Cool and strain.
3. Spritz on hair and expose to sunlight.

Turmeric Scalp Soother

Region of Use: Ayurveda

Functional Benefits: Antimicrobial, anti-inflammatory, soothes irritation

Materials:
- 1 tsp turmeric powder
- 1 tbsp aloe gel
- Optional: 1 drop tea tree essential oil

Protocol:
1. Mix thoroughly.
2. Apply to affected areas.
3. Leave for 15 minutes. Rinse.

Moringa Infusion

Region of Use: African traditional medicine

Functional Benefits: Promotes hair strength and nourishment

Protocol:
1. Steep moringa leaves or powder in warm oil.
2. Apply as a hot oil treatment.

Baobab Butter Balm

Region of Use: African natural skincare

Functional Benefits: Rich in omega fatty acids; emollient for scalp and hair

Materials:
- 1 tbsp baobab oil or butter
- 1 tbsp shea butter
- 5 drops lavender essential oil

Protocol:
1. Melt oils together gently.
2. Apply to scalp overnight or use as leave-in balm.

These formulations serve as more than just cosmetic tools; they represent an intergenerational inheritance of botanical science and cultural reverence. May you practice with intention, care, and continuity.

Zero-Waste Beauty Living

Reuse, Upcycle, Reduce, Reconnect

In a world bursting with throwaway packaging and chemical-laden routines, choosing a zero-waste path is more than eco-friendly—it's revolutionary. It's a quiet rebellion against the "more-is-more" mindset. It's choosing intention over impulse, ritual over routine, and connection over consumption.

But here's the truth: you don't need to be perfect. You don't need to make all your own mascara or swear off packaging forever. Zero-waste beauty isn't about guilt. It's about awareness. Small changes. Beautiful steps.

Whether you live in a house, a van, a tent, or a tiny off-grid haven, this offers practical ideas and soulful shifts to simplify your beauty world—and lighten your footprint while deepening your roots.

Why It Matters

- Most mainstream beauty products come in plastic—much of which ends up in landfills or oceans.
- Your skin absorbs what you put on it—chemicals, preservatives, fragrances.
- The Earth offers powerful alternatives: herbs, oils, clays, and water.

Choosing zero-waste is choosing to trust the Earth again. And to trust yourself.

The Core Pillars of Zero-Waste Beauty

1. Reuse

- Wash and repurpose old jars, dropper bottles, and tins.
- Keep tiny spoonfuls of herbs or powders for travel kits.
- Repurpose old face cloths into reusable makeup wipes or compress cloths.

2. Upcycle

- Use coffee grounds from your morning brew in a body scrub.
- Infuse citrus peels in vinegar to clean your space.
- Turn cardboard into DIY soap labels or gift tags.

3. Reduce

- Simplify your beauty shelf: less products, more potency.
- Skip the "one for every problem" approach. Instead, find one balm that heals many things
- Choose refill stations when available, or buy in bulk and share with a friend.

4. Reconnect

- Grow your own ingredients where possible—even if it's a single pot of rosemary.
- Know your makers. Support small batch, handmade, ethical brands.
- Remember: everything you use should be able to return to the earth.

Zero-Waste Tips for All Living Setups
In a Tiny Home or Caravan:
- Use stackable containers to save space.
- Dry herbs by hanging them from string or window racks.
- Compost in small batches with bokashi buckets or worm tubes.

In the Garden or Rural Home:
- Grow herbs you regularly use: mint, calendula, thyme, aloe.
- Use rainwater for rinsing and infusions.
- Share homemade goods in reused jars with neighbours or barter for other local items.

In a Rental or Shared House:
- Store dry ingredients in labeled spice jars.
- Skip the plastic loofah—try a hemp cloth, washcloth, or konjac sponge.
- Keep a beauty ritual tray you can move between spaces.

Beauty Without Waste Looks Like This…
- A bar of shampoo wrapped in paper.
- A single oil that serves as cleanser, moisturizer, and balm.
- Dried herbs in jars labeled with your handwriting.
- A compost bin that smells like citrus and mint.

A bathroom shelf that feels like an altar, not a stockpile.

Start Here: Easy Swaps

If you're beginning your journey into low-waste beauty, start with a few simple yet powerful swaps. Replace your liquid shampoo bottles with a solid shampoo bar, and trade disposable razors for a long-lasting stainless steel safety razor. Instead of single-use cotton rounds, opt for reusable cloth wipes that can be washed and reused. Switch out your plastic toothbrush for a bamboo toothbrush, and say goodbye to plastic-wrapped soap by choosing homemade or package-free options. Finally, instead of using bottled toner, try a herbal face mist in a reusable glass spray bottle. These small changes, when practiced consistently, ripple out into big impact—not only for the Earth but for your everyday ritual.

You don't have to do it all today. Maybe your first step is refusing one more plastic bottle. Or making your own face oil. Or reusing a single jar.

Each act is a prayer. A promise.

To the Earth. To your future self. To the generations who will garden in your footsteps.

A Living Legacy

Teaching Others, Sharing Wisdom, Preserving Culture
There is a moment in every gardener's life—quiet and unexpected—when the focus shifts from what you can grow to what you can give. When your hands no longer dig just for harvest, but for heritage. When your knowledge becomes less about personal healing, and more about passing the flame.
That moment is the seed of legacy.
And in a world that often forgets the wisdom of the Earth, your choice to remember—to share, to teach, to preserve—becomes an act of revolution.
You don't need to be an expert. You don't need a certificate. You need only a lived experience, a curious heart, and a willingness to speak from the soil of your truth.

You Are a Knowledge Keeper Now
Every balm you've made. Every seed you've planted. Every herbal rinse or moon bath. These are now part of your story. And stories are meant to be passed on.
Whether you're a parent, an aunty, a big sister, a friend, a wandering teacher, or a quiet example—someone is watching. Learning. Absorbing the energy of your choices.
Let them.
Teach by living well.

Preserving Culture Through Plants

In every country, every culture, every lineage—plants hold stories. They are the original record keepers.
When you grow with intention, you honour more than the Earth—you honour those who came before you. Especially if you carry Indigenous, ancestral, or marginalised roots, your garden becomes a living archive. Each calendula petal. Each sage leaf. Each root pulled from the soil—these are your heirlooms.
So teach the names in your language. Pass on the rituals. Celebrate the old ways and adapt them for the now.
You are not just growing plants.
You are reclaiming what was almost lost.

What Legacy Really Looks Like
It's not a statue or a perfect book or a flawless garden.
It's:
- A niece who remembers how to make lavender mist.

- A friend who grows mint on their windowsill because of you.
- A child who calls yarrow "the cut-stopper plant."
- A neighbour who composts now because you told them it mattered.
- A stranger who read your words and started saving seeds.

That's legacy. Quiet. Real. Rooted.

The Path Forward Is Circular

You may feel like you're just one person. But in the garden of life, even the smallest seeds ripple outward. One shared salve becomes five. One taught ritual becomes memory. One raised garden bed becomes a community.

And when you're gone, may the Earth remember you by the herbs that still grow wild where you once walked.

Reflections from the Garden

There is a sacred silence that comes after planting. After harvesting. After making a balm or rinsing your hair with herbs. It's the hush where integration happens. Where you pause—not to do more, but to feel what's already growing inside you..

You've learned how to garden with reverence. To craft with plants. To honour culture. To tend beauty with the hands of a healer.

Now, let's slow the rhythm. Let's listen inward.

These prompts, affirmations, and meditations are like sitting barefoot in the garden at dusk—where the real growth happens beneath the surface.

Seed Journal

DATE / /

S M T W T F S

PLANT NAME ....

HARVEST DATE ....
GERMINATION DATE ....
FIRST LEAVES ....
FLOWERING ....

LOCATION IN GARDEN

....
....

COMPANION PLANTS

....
....

NOTES

....
....
....
....

Seed Journal

DATE   /   /

S M T W T F S

PLANT NAME ....

HARVEST DATE ....
GERMINATION DATE ....
FIRST LEAVES ....
FLOWERING ....

LOCATION IN GARDEN

....
....

COMPANION PLANTS

....
....

NOTES

....
....
....
....

Seed Journal

DATE   /   /

● ● ● ● ● ●
S M T W T F S

PLANT NAME ....

HARVEST DATE ....
GERMINATION DATE ....
FIRST LEAVES ....
FLOWERING ....

LOCATION IN GARDEN

....
....

COMPANION PLANTS

....
....

NOTES

....
....
....
....

Seed Journal

DATE / /

S M T W T F S

PLANT NAME:

HARVEST DATE
GERMINATION DATE
FIRST LEAVES
FLOWERING

LOCATION IN GARDEN

COMPANION PLANTS

NOTES

Seed Journal

DATE   /   /

● ● ● ● ● ● ●
S M T W T F S

PLANT NAME ....

HARVEST DATE ....
GERMINATION DATE ....
FIRST LEAVES ....
FLOWERING ....

LOCATION IN GARDEN

....
....

COMPANION PLANTS

....
....

NOTES

....
....
....

DATE CREATED: /   /

● ● ● ● ● ●
S   M   T   W   T   F   S

RECIPE NAME

INGREDIENTS USED   METHOD

RESULTS

NOTES

DATE CREATED   /   /

● ● ● ● ● ●
S M T W T F S

RECIPE NAME

INGREDIENTS USED   METHOD

RESULTS

NOTES

Let This Be the Beginning

If this book has rested in your hands, on your lap, or beside your seed packets and tea mug—thank you. Thank you for walking barefoot through these pages with me.

You have learned the rhythm of the garden. You've stirred petals into potion. You've remembered that the Earth is not a stranger, but a mother, a mirror, a teacher.

But this is not the end.

This is the moment between breaths—between bloom and seed, between compost and creation. This is the fertile pause. The quiet before the next planting.

So take what you've gathered: The recipes.

The rituals.

The reflections.

The remembering.

And let them grow in your life—not just in jars and beds, but in your choices, your care, your community.

May your hands always know how to mix oil with flowers.

May your feet always find the path back to the garden.

May your voice be one that tells stories of calendula, and moonlight, and the old ways made new.

And when someone asks where you learned all this—

you can say, with a soft smile,

"From the Earth. She wrote it on my skin."

With muddy feet and a full heart,

Cailin Cooper

Your companion in soil, salves, and sacred ritual

www.ingramcontent.com/pod-product-compliance
Lightning Source LLC
Chambersburg PA
CBHW042128100526
44587CB00026B/4212